"*Posthuman research playspaces* is a creative, future facing, and politically engaged work that takes the reader on a journey into a different relationship with our environment. If we want to stop climate change, then how would we need to think? What would we need to do? Taking cues from children, the environment and philosophy, Rousell and Cutter-Mackenzie-Knowles offer responses to these questions that carry their reader into play spaces where the materiality of nature is a teacher. This book will be essential reading for those in environmental education, sustainability studies, childhood studies, arts education, and the posthumanities".

Professor Anna Hickey-Moody, *School of Media and Communication, RMIT University, Australia*

"*Posthuman research playspaces* is a unique book that is engaging, challenging and ultimately hopeful in addressing climate change and its implications for children's lives. The book leaves no stone unturned, calling our attention to issues of colonisation that are intimately entangled with the destruction of planet Earth. Putting posthumanist theories into conversation with children's art, media, and writing, the authors offer detailed analyses of creative methods which extend children's capacities to inquire and express insights into climate change. I love this book and imagine all readers concerned with childhood and climate change will love it too".

Professor Margaret Somerville, *School of Education, Western Sydney University, Australia*

POSTHUMAN RESEARCH PLAYSPACES

Posthuman research playspaces: Climate child imaginaries addresses the need for new forms of climate change education that are responsive to the rapidly changing material conditions of children's socioecological worlds.

The book provides a comprehensive understanding of how posthumanist concepts and methods can be creatively developed and deployed in collaboration with children and young people. It connects climate change education with posthumanist studies of childhood in the social sciences and environmental humanities. It also offers opportunities for readers to encounter new theoretical and methodological approaches for collaborative art, inquiry, and learning with children. Drawing on three years of participatory research undertaken with 135 children in the Climate Change and Me (CC+Me) project, it takes children's creative and affective responses to climate change as the starting point for the co-production of knowledge, community engagement, and the transformation of pedagogy and curriculum in schools. Thinking through process philosophy, and in particular, the works of Whitehead and Deleuze, the book develops new concepts and methods of creative inquiry which situate children's learning, aesthetic production, and theory-building within a more-than-human ecology of experience.

The book presents a series of generative openings and propositions for future research in the field of climate change education, while also offering wide-ranging applications for graduate students and researchers in childhood and youth studies, the environmental arts and humanities, cultural studies of science and technology, educational philosophy, and environmental education.

David Rousell is Senior Lecturer in Creative Education at RMIT University, Australia, where he co-leads the Creative Agency Research Lab for ecological studies of creativity. His book *Immersive Cartography and Post-Qualitative Inquiry* was published with Routledge in 2021.

Amy Cutter-Mackenzie-Knowles is Professor of Sustainability, Environment and Education at Southern Cross University, Australia. She is the Executive Dean of the Faculty of Education, as well as the Research Leader of the "Sustainability, Environment and the Arts in Education" (SEAE) Research Cluster.

POSTHUMAN RESEARCH PLAYSPACES

Climate child *imaginaries*

David Rousell and
Amy Cutter-Mackenzie-Knowles

Routledge
Taylor & Francis Group
LONDON AND NEW YORK

Designed cover image: Sam Lucena

First published 2023
by Routledge
4 Park Square, Milton Park, Abingdon, Oxon OX14 4RN

and by Routledge
605 Third Avenue, New York, NY 10158

Routledge is an imprint of the Taylor & Francis Group, an informa business

British Library Cataloguing-in-Publication Data
A catalogue record for this book is available from the British Library

ISBN: 978-1-032-37233-4 (hbk)
ISBN: 978-1-032-37238-9 (pbk)
ISBN: 978-1-003-33600-6 (ebk)

DOI: 10.4324/9781003336006

Typeset in Bembo
by SPi Technologies India Pvt Ltd (Straive)

Access the Support Material: www.routledge.com/9781032372389

For Forest, River, Lily & Finley

CONTENTS

FOREWORD

12-year-old Climate Change and Me researcher

Is my occurrence on the face of this planet just to create a life that may support me... or to create a life for others too?

It appears that we humans have lodged an idea into our minds that we must fulfil every need of ours; every craving, desire, anything that could help to increase our over-enlarged egos. This is the main reason why climate change has been worsening. Our species has developed a sophisticated society, where each new generation paints many new ideas and lifts the rate of environmental destruction. This continually reduces the natural world; a world which we cannot live without; a world that's far away from today's high-tech gadgets and devices.

Sadly, we are virtually blind to our destructive habits; actions that are centred at the expense of our own well-being, and the flora and fauna that inhabit the Earth. Always wishing for bigger ... better ... more; so much so that we never appear to be satisfied or pleased. We find ourselves currently facing a global climatic crisis due to each and every one of us, where each new thing is only admired for a short period of time before we establish a new desire; a cycle which repeats continually. We must change our values and what we believe is important to us – this change mustn't only occur in a small minority of the population, but almost all seven billion of the humans who inhabit Earth.

Our lifestyles directly affect not only us humans, but the lives of many diverse species of animals, plants, and bacteria too. Life as we know it must change for the health and well-being of the future. Every creature has a life centre and needs fresh air and water to remain alive. As humans, we may think of ourselves as being superior and better than all other life forms due to our advances in the creation of "technology". By stepping forward and learning how to make our lives better and outliving many, we've been progressively destroying the habitats of many; causing species' extinction rates to rise to unprecedented levels in history.

The health of the planet is directly correlated with our own health. We are so wrapped up in what's happening in modern, heavily materialistic lifestyles (the focus upon sports, fashion and the media, etc.), that we have just evolved to disregard/not care for what really does matter. We are in a routine of selfishness and consuming more, and more. We have missed out on the basics of nature, our connection to it, and its role in our well-being and how it's essential for our survival. We've been destroying animals' lives and their habitats daily without even realising the downfall to our selfishness. We must prevent worse disasters that ruin life as they know it and have the potential to ruin life as we, ourselves, know it.

Many people are still unaware of the steps they must take and responsibility in reducing their carbon outlet, and reducing their overall impact. The only possible solution is if humankind realises that we must evolve to become more realistic with our lifestyles, opinions, focuses, and choices; we must take action towards regenerating our tainted planet, before the thoughtless destruction of environmental resources has gone so far that there can be no reversal. This issue must become the top priority within every single nation. We must band together to reverse the effects of climate change which we, ourselves have caused.

The difference must begin with us. We must make drastic changes to how we think and make choices about our everyday activities and focuses. We were not placed on this Earth to make an acquisitive and ideal life that supports the human race only; we were placed on this planet to foresee a life for all things, whether they are living, or non-living, plant or animal. Not to destroy the climate by means of our own self-centred desires.

Nikki Whitehead

ACKNOWLEDGEMENTS

With permission, across this book we draw upon concepts, theories, and data previously published in the following works:

Cutter-Mackenzie, A., & Rousell, D. (2019). Education for what? Shaping the field of climate change education with children and young people as co-researchers. *Children's Geographies*, *17*(1), 90–104. https://doi.org/10.1080/14733285.2018.1467556

Cutter-Mackenzie-Knowles, A., & Rousell, D. (2020). The mesh of playing, theorising and researching in the reality of climate change: Children's chldhoodnature research playspaces. In A. Cutter-Mackenzie-Knowles, K. Malone, & E. Barratt Hacking (Eds.), *Research handbook on childhoodnature: Assemblages of childhood and nature research* (pp. 199–222). Springer Nature.

Rousell, D., & Cutter-Mackenzie, A. (2015). *The Changes – Art, writing and research by student researchers in the climate change and me project.* NSW Environmental Trust.

Rousell, D., Cutter-Mackenzie, A., & Foster, J. (2017). Children of an earth to come: Speculative fiction, geophilosophy and climate change education research. Special Issue for *Educational Studies*, 53(6), 654–669. https://doi.org/10.1080/00131946.2017.1369086.

Rousell, D., & Cutter-Mackenzie-Knowles, A. (2020a). A systematic review of climate change education: giving children and young people a 'voice' and a 'hand' in redressing climate change. *Children's Geographies*, *18*(2), 191–208. https://doi.org/10.1080/14733285.2019.1614532.

Rousell, D., & Cutter-Mackenzie-Knowles, A. (2020b). Uncommon worlds: Towards an ecological aesthetics of childhood in the Anthropocene In A. Cutter-Mackenzie-Knowles, K. Malone, & E. Barratt Hacking (Eds.), *Research handbook on childhoodnature: Assemblages of childhood and nature research* (pp. 1657–1680). Springer Nature.

We also take this opportunity to thank Katie Hotko for her tireless assistance with copyediting.

We especially thank the young researchers who made this book and scholarship possible. They generously afforded their time, intellect, and passion. Their research undoubtedly contributes to a posthuman earth to come.

And finally, thank you to our families who unconditionally support the work we do with love and care. We dedicate this book to them, and our children in particular: Forest, River, Lily, and Finley.

1

ENCOUNTERING THE ANTHROPOCENE

Introduction

Today's children and young people are the first to have lived their entire lives under the known existential threat of global climate change. For many young people climate change has become the defining political issue of their lives, with many compelled to take urgent action on a life-threatening ecological crisis which previous generations have lacked the political will, organisation and vision to address. On March 15, 2019 an estimated 1.4 million children and young people in over 150 countries left school to protest against governmental and corporate climate inaction as part of the YouthStrike4Change movement. The movement's leader, 15-year-old Swedish activist Greta Thunberg, called on the young people of her generation "to hold the older generations accountable for the mess they have created, and expect us to live with" (Thunberg, 2019a).

The recent YouthStrike4Change movement, along with Extinction Rebellion and numerous other youth climate activism groups, has brought unprecedented public attention to the impacts of climate change on children and young people's lives. These youth-led movements are publicly reframing climate change as a profound issue of generational, social, and environmental injustice, emphasising the failure of humanist ethical ideals and techno-scientific regimes in securing a liveable future for the majority of human and nonhuman inhabitants of the planet (O'Brien et al., 2018; Ritchie, 2021). For many scholars in the social sciences and humanities, these changing material conditions of Earthly life have taken on the contested name of "Anthropocene", a new epoch defined by the geologic scale of human impact on the ecological functioning and trajectory of the planet as a whole (Ellsworth and Kruse, 2012; Lewis and Maslin, 2015; Morton, 2016). While the Anthropocene concept has now been widely adopted as both a scientific and cultural descriptor of the current era, it is clear that both the causes and effects of the

DOI: 10.4324/9781003336006-1

Anthropocene are by no means equally distributed (Steffen et al., 2015), with many of those least responsible for the onset of climate change often positioned as most vulnerable to its impacts (Baldwin and Bettini, 2017; Taylor, 2017).

The international youth climate protests of 2019 could be considered the first mass mobilisations of an "Anthropocene generation" which refuses to participate in a society and an education that fails to take ethical responsibility for the crisis of climate change. Turning the tables on the humanist image of childhood as the biological and cultural "future of humanity" (Lee, 2013; Sheldon, 2016), these youth movements are shaping alternative climate change imaginaries from a position of immanent eco-critique and refusal. Two weeks after the mass climate strikes of March 15th, Thunberg addressed members of Parliament in the UK with the following message:

> You lied to us. You gave us false hope. You told us that the future was something to look forward to. And the saddest thing is that most children are not even aware of the fate that awaits us. We will not understand it until it's too late. And yet we are the lucky ones. Those who will be affected the hardest are already suffering the consequences. But their voices are not heard.
>
> *(Thunberg, 2019b)*

We commenced writing this book in the summer of 2019, as young people orchestrated global climate strikes calling on older generations to join them in refusing to attend work, university, or school in protest against climate inaction (FridaysForFuture, 2019). As we completed the book in 2021, we found ourselves in the midst of a global pandemic that has severely altered the very futures that today's children and young people were expected to fulfil.

Climate Change and Me (CC+Me)

In 2014, six years before the #ClimateStrike4Action movement captured public attention, we began a project called Climate Change and Me (CC+Me) in the Northern region of NSW, Australia. The CC+Me project took shape as a three-year collaboration with 135 children and young people (aged 9–14) investigating the experiential impacts of climate change in their everyday lives. The project was open-ended and directly responsive to the emerging interests and sensibilities of young people. Children and young people were offered a platform to develop their own modes of creative inquiry into climate change in their communities. We followed their trajectories of inquiry as entry points for philosophical experimentation and creative co-production, leading us to explore the potentials of environmental art, film-making, poetics, philosophy, and fiction in developing alternative modes of climate change education. To our knowledge, this was the first large-scale participatory project to tackle children's experiences of climate change from a posthumanist, arts-based perspective, pivoting on the contested concept of the Anthropocene to engage with climate change outside of conventional nature/culture binaries.

When we started this project in 2014, very little was known about what children and young people thought about climate change, how it was affecting their everyday lives, or how they might address the issue through their own concepts and experiences. Climate change and childhood were widely discussed in separate spheres of discursive deliberation and engagement. If climate change was addressed at all in Australian schools, it was largely relegated to a minor topic in the high school science curriculum and often delivered as a contested topic of scientific debate (c.f. McNeill and Vaughn, 2012; Shepardson et al., 2012). Our systematic review showed that research on climate change education had primarily focused on young people's erroneous scientific knowledge and lack of "factual" information, largely ignoring the profound social and affective dimensions of children and young people's lived realities of inheriting and inhabiting an age of climate change (Rousell and Cutter-Mackenzie-Knowles, 2020). We realised that the majority of young people had been cast adrift when it came to climate change, left to simply grapple with the problem alone. We were witnessing the unfolding of a planetary-scale existential threat that was casting an invisible weight over children's lives, and yet there was often nowhere for them to explore questions about climate change, or express feelings, thoughts, and concerns about the issue.

The CC+Me project responded to the need for open spaces of inquiry in which children and young people could encounter climate change, and experiment creatively with different possibilities for sensing, attuning, and responding to its complexity. This book explores alternative forms of climate change education research driven by the experiences, practices, and imaginaries of the 135 Australian children and young people who participated in the CC+Me project. The book is animated by two central questions:

> How can climate change education become more attuned and responsive to the changing conditions of children and young people's lives in the Anthropocene?
>
> How might children and young people contribute to a critical re-imagining of climate change education, in ways that open up new imaginaries and potentials for future co-existence?

The *Climate Change and Me* project addressed these questions through five distinct phases of creative, participatory, and applied research (Figure 1.1). The first phase of the project involved the formation of an open-ended research collective through a series of collaborative workshops in schools and Southern Cross University. Through these workshops, we established an intergenerational research collective that included ourselves (two adult researchers) and approximately 135 children and young people between the ages of 9 and 14. These workshops introduced children and young people to a range of scientific, social, and political perspectives on climate change, and the emergence of the Anthropocene as a figuration of the current geosocial epoch. Children and young people also learned about practices of qualitative and arts-based research in education, with workshops dedicated

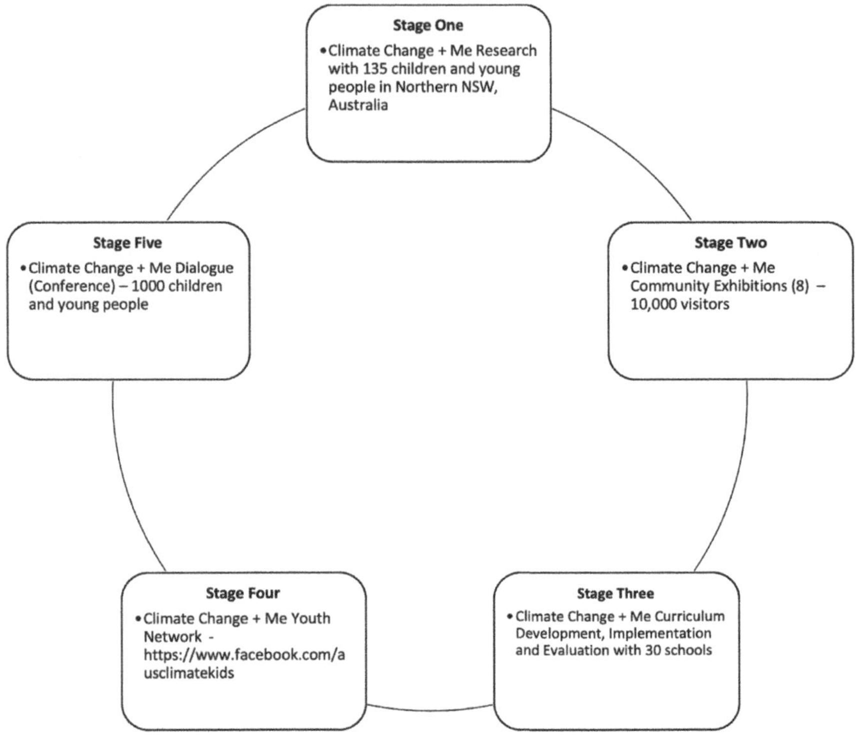

FIGURE 1.1 Key phases of the Climate Change + Me project.

to developing creative techniques of inquiry into the impacts of climate change within their communities and environments. Questions of ontology, epistemology, ethics, and aesthetics were explored through collaborative thought experiments, often utilising iPads as mobile research tools for developing techniques of collective inquiry.

These initial provocations and creative workshop activities were not prescriptive but *propositional*, in Whitehead's (1978, p. 254) sense of activating propositional feelings towards novel contrasts and tonalities of experience. The workshops did not seek to "train" children how to do research but, rather, to generate a space for the mutual activation of propositional feelings along novel trajectories. In Chapter 3 of this book, we theorise this methodological approach to working with children as the activation of "posthuman research playspaces". By following the emergent contours of these research playspaces, we saw ourselves and the young researchers who joined us on the project continuously and fluidly shifting roles, as the playspaces refused to settle into categorical regimes of leadership, authority, or control over what happened.

In between workshops, the research collective continued to share their evolving trajectories of inquiry and research artefacts through a dedicated social media space on the project's website. This enabled members of the collective to share and comment on photographs, videos, interviews, sound recordings, poems, stories, imaginings, and even games or challenges that emerged spontaneously through the online platform.

Following this initial period of intensive research, we organised another series of workshops called "Thinktanks". These workshops were designed around collaborative processes for curating, analysing, and theorising the wealth of images, sound and video recordings, writings, and conversations that had been co-created by the intergenerational collective. These processes were primarily arts-based, and were designed to feed into the travelling *Past Now Future* exhibition that took place in the second phase of the project. Importantly, the shared goal of preparing collaborative work for public exhibition became a powerful catalyst for children and young people's participation in the project, as it enabled them to directly speak back to their own communities in creative and participatory ways. The *Past Now Future* exhibition was composed of three works: *Pieces of the Earth*, a manipulable mosaic of 200 images suspended in encaustic wax medium; *Voices from the Anthropocene*, a series of interactive video installations featuring films co-created by children and young people; and *The Changes*, a printed book featuring critical writings, interviews, photo essays, poetry, drawings, and speculative narratives composed by children and young people. The exhibition travelled to eight public libraries and gallery spaces across Northern NSW in 2015, and was viewed by over 10,000 members of the public.

In the third phase of the project, we drew on the extensive research and curatorial analysis completed in the earlier phases as the basis for co-designing a climate change curriculum. This involved the assembly of a curriculum development team of sixteen members, including five children and young people from the project, five school teachers, three lecturers in education, a research assistant, and the two of us. The design of the curriculum was based on the collective inquiry process that children and young people had generated through their involvement in the first phases of the project. By following how children and young people developed their own trajectories of inquiry through creative practices, we were able to map an open-ended and transdisciplinary inquiry process onto the English, Visual Arts, Geography, and Science key learning areas in the Australian National Curriculum. The works of art, films, poetry, essays, and speculative fiction created by children and young people in the first phases also became the core resources, perhaps even "textbook", for the climate change curriculum. This resulted in a curriculum constructed almost entirely from children and young people's creative practices and ideas. It has since been taken up in over 25 schools across Australia.

The fourth and fifth phases of the project involved a series of outreach and social impact programs in both digital and physical spaces. The fourth phase involved the co-creation of a digital network that expanded the CC+Me web forum into a wider national collective through social media. This included the founding of the Australian Climate Kids Network on Facebook, and the cultivation of relationships with existing youth climate activism groups, such as the Australian Youth Climate Coalition and SEED Indigenous Youth Climate Network. The fifth phase of the project pivoted on a series of trials and evaluations of the CC+Me curriculum in five schools across New South Wales. These schools were then invited to attend an event we called the Climate Change Challenge, along with the six schools who had initially participated in the earlier research phases. The Climate Change Challenge was the final large-scale event of the project, which took shape as a kind of "take-over" of Southern Cross University's Lismore Campus by over 1,000 children and young people. Organised much like a small festival, the Challenge included indoor and outdoor workshops, video installations, organic food, a climate change photo competition, climate art exhibition, and speaking events led and coordinated by children and young people.

As we write this book six years after beginning CC+Me, we are continuously amazed by the prescient intensity of the research that children and young people cultivated and expressed through the project. Looking back on the extraordinary proliferation of events, artefacts, data, and experiences that made the project what it was/is, we can sense the stirrings of a much larger and more widely distributed youth movement just beginning to come to expression. Our approach to writing this book seeks to foreground this sense of the incipient and the germinal, of minor movements and intensities which are always just in the process of coming to expression (Manning and Massumi, 2014). Rather than attempting to report upon or document the project as a "complete" or static body of work, we are interested in following minor variations of feeling and thought that children and young people injected into the research assemblage, propositions that continue to resonate and resound in ongoing youth struggles for environmental and social justice in the passing present.

Climate Child *imaginaries*

Throughout this book, we engage with the concept of the "climate child imaginary" as an open-ended figuration of the material, affective, and social conditions through which an age of climate change is collectively experienced and entangled with childhood. The concept of a social imaginary has been widely taken up in the social sciences for several decades. It has been developed in Appadurai's (1996) anthropology of global transcultural flows, Castoriadis' (1987) account of symbolic world-making through collective agency, and Williams' (1977) notion of "structures of feeling" that condition a shared affectivity or "feeling" of a historical epoch. The sensorial and embodied nature of the social imagination is further emphasised by Hickey-Moody et al., (2016), who draw on Spinoza's philosophy to connect the social imaginary to our "sensory experience of the world, an experience that is a kind of collective awareness, as our bodies are an extension of our contexts" (p. 130). They develop the concept of "feeling futures" to describe the role of the imaginary in the everyday lives and educational aspirations of disadvantaged young people. Building on Gatens and Lloyd's (1999) feminist reengagement with Spinozan philosophy, they describe how a collective imaginary is felt intensively at the bodily level, while, at the same time, participating in complex relational affinities with other bodies.

> One of the effects of Spinoza's understanding of the mind and body as individually specific and culturally grounded is that an individual's imaginative

> associations are always considered products of the shared affinities of human bodies. One's imagination is, by definition, the product of one's relationships with other bodies. Thus, one person's beliefs and actions directly limit or extend the actions and beliefs that other people associated with it can undertake.
>
> *(Hickey-Moody et al., 2016, p. 131)*

This engagement with the embodied imagination as an individuating and collective "feeling" of climate change establishes a running theme through this book. This entails new intersections and contributions of our work in relation to contemporary affect studies, an increasingly divergent and capacious collection of studies often arranged under the subsuming heading of "affect theory" (Seigworth and Gregg, 2010). Spinozan and Deleuzian readings of affect form a robust, yet contested, genealogy within this expansive body of work (Bertelson and Murphie, 2010; Hickey-Moody, 2013; Manning, 2016; Massumi, 2011), which also draws on a broad variety of psychoanalytic (Stern, 2010; Winnicott, 1989), neuroscientific (Papoulias and Callard, 2010), and cultural interpretations of affect (Berlant, 2011; Williams, 1977). Definitional and operational accounts of affect differ significantly across these varied instantiations, ranging from transindividual, more-than-human accounts of impersonal forces and intensities (c.f. Massumi, 2018) to personalised narratives of the sociability of emotional states (c.f. Ahmed, 2008). Our particular intercession into these debates pivots primarily on the former, process-orientated approaches that build on the Spinozan, Whiteheadian, and Deleuzian systems of non-anthropocentric thought. Specifically, we turn to Whitehead's speculative empiricism to develop the concept of the climate child imaginary as a shifting and dynamic ecology of affects produced at the intersections of climate change and childhood experience.

Our entry point into the study of the climate child imaginary as an affective ecology can be described as resolutely *aesthetic* in orientation. We understand the notion of a climate child imaginary in the sense of an aesthetic *figuration*, as posed by Haraway (2016), to describe a living entanglement of embodied and incorporeal forces that compose not only "what is", but also "what may be". For Haraway, an aesthetic figuration like the climate child imaginary is capable of entangling a vast range of material, affective, and social "facts", ranging from climate science to science fiction, digital worlds, Indigenous mythology, and creaturely kinships. In this respect, the book's speculative focus on climate change imaginaries does not constitute a rejection of "science" as such, but rather, an affirmation that *many other sciences are possible* when diverse cosmologies and ecologies of practices are taken into account (Cajete, 2006; Rousell et al., 2022; Stengers, 2018). This means that our aesthetic sense of a climate child imaginary is shifting continuously, and never coalesces into a universal figure or transcendent model. The imaginary shifts with each experiential encounter with the world, as a rapidly mutating confluence of matters of ecological fact, value, care, feeling, thought, and concern.

And yet, as a collectively sensed feeling and gathering of intensity, aesthetic figurations of the climate child imaginary are no less actionable in their ambiguity and incalculability. As cultural theorist Lauren Berlant (2011) argues, an "aesthetics of the affective event" can "make certain situations exemplary laboratories for sensing contemporary life in new idioms of affective realism" (p. 54). In our case, these exemplary laboratories are the situations of children's affective encounters with climate change in their everyday lives and more-than-human environments.

Throughout this book, we pursue an affective realism of the climate child imaginary through the diverse perspectives of children and young people, pivoting on their collective and individual aesthetic practices and embodied knowledges of climate change in the current epoch. While our study of climate child imaginaries attends to the ordinary, everyday intimacies of children's experiences of climate change through their creative practices, it also foregrounds the power of speculative thought in cultivating a metaphysics of childhood and nature that is responsive to environmental transmutations at planetary scale. We draw extensively on the work of Whitehead and Deleuze as speculative philosophers who help us think-with children and young people about what a climate child imaginary can be(come). For Whitehead (1968, 1978), speculative philosophy begins with our everyday capacities to sense, feel, think, and experience the world, before taking flight into the "thin air" of the imagination in attempting to account for the metaphysical *conditions* that make such experience possible. On returning from the flight of the speculative imagination, we become capable of grasping new knowledge and insights about ourselves and the world that combine the empirical with the rational, the sensorial with the imagined, the faculties of reason with visceral bodily feeling.

The role of the speculative, for Whitehead, is ultimately one of generalisation with respect to the shared conditions of experience, or "common world" which also constitutes the massively distributed habitat of a cosmic epoch. Whitehead asks: What do everyday bodily practices of sensing, thinking, inhabiting, imagining, relating, and caring disclose about the metaphysical conditions of this common world? What are the general concepts capable of accounting for the affective *conditions* of this common world of experience? Whitehead asks these speculative questions not only of humans, but of nature, and as such, all things that exist. As we elaborate in Chapters 4 and 5, Whitehead's body of work offers a system of thought that operates entirely through affective feeling and aesthetic experience, a non-anthropocentric ecology which grants all things and events the capacity to affect, and be affected by, the ever-changing "common world" of nature as an immanent totality. Whitehead uses the term "feeling" to describe the aesthetic ontogenesis (or coming into being) of experience through the affectivity of relational events, a process of becoming that is always both physical and mental, ecological and aesthetic, objectively realised and subjectively felt. There is, in this account, no formal distinction between human thought and other creaturely permutations of feeling which grasp and take account of the world in a teeming variety of ways.

Throughout this book, we elaborate, hone, sharpen, and repurpose Whitehead's speculative concepts as philosophical tools for exploring a multiplicity of climate child imaginaries with children and young people, including the concepts of nature, feeling, prehension, creature, becoming, and worldly sensibility. For instance, by taking up the concept "feeling" in the Whiteheadian sense, our accounts of childhood experiences of climate change are by no means bounded by the phenomenological container of the human-body-mind complex, and are therefore seen to extend and intensify across all scales and aesthetic orders of existence (atomic, chemical, cellular, organismic, planetary, and so on). This gives our book a distinctive take on both the affective and posthumanist turns in education and the social sciences more broadly, contributing to a dynamic resurgence of interest in Whitehead's philosophy to address questions of more-than-human affectivity, sociality, and learning under current conditions (c.f. Atkinson, 2017; Cole and Mirzaei Rafe, 2017; Cole and Somerville, 2017; Mazzei, 2021; Rousell, 2021).

Alongside this theoretical engagement with process philosophy, our study and cultivation of climate child imaginaries involve an emphasis on creative practices as aesthetic modes of participatory research and activism capable of generating alternative climate futures. Our research provides evidence that art can mobilise children and young people's responses to the challenges of climate change "through aesthetic modes of address which are not confined by an adherence to moral, scientific or discursive regimes of expression" (Turpin and Davis, 2015, p. 4). The creative practices of children and young people demonstrate how art does not function to refute or eschew logic, science, morality, or discourse but, rather, creates a new aesthetic milieu in which the world might be encountered differently, on terms that are affective, axiological, and onto-epistemological, rather than cognitive and logocentric (Shaviro, 2015). In this respect the CC+Me project can be understood as an experiment in working collectively with children and young people to cultivate new arts of living and learning on a damaged planet (Demos, 2020; Tsing, 2015) through an affective realism of aesthetic figurations (Berlant, 2011). We propose that speculative and artistic modes of inquiry can offer distinctive platforms for children to participate in the development of new climate imaginaries steeped in the cultures of the minor (Manning, 2016), the critical posthuman (Braidotti, 2013), the nomadic (Deleuze and Guattari, 1987), the micropolitical (Rolnik, 2017), and the decolonial (Yusoff, 2018). Crucial to this book is the reconsideration of children and young people as researchers, as artists, writers, scientists, and thinkers who are intimately attuned to the lived dynamics of climate change and other critical issues of social and environmental justice. We offer this work as a distinct call for children and young people to actively reshape the very possibilities of climate change education through critical speculation and the co-creation of alternative social imaginaries.

Encountering the "Anthropocene"

> We kill many things. We are malicious killers.
>
> **Nikki, age 12, CC+Me Researcher**

Global climate change is one of numerous planetary-scale transmutations currently reconfiguring the very nature(s) of life in the contemporary moment. Children and young people are growing up in an age when the human population has tripled in a single lifetime (McNeill and Engelke, 2014), while 52% of the world's nonhuman animal life has been decimated over the same period (World Wildlife Fund, 2014). Digital sensing, communication, and computational technologies are now pervasive elements of social and ecological systems worldwide, transforming the Earth into a "computational planet" (Gabrys, 2016). The combined effects of anthropogenic climate change, mass extinction, genetic manipulation, ubiquitous computing, and synthetic biology are eroding conventional boundaries between human and nonhuman, life and nonlife, the natural and the artificial (Bignall and Braidotti, 2019). While the "Anthropocene" concept emerged from the Earth sciences in the early 2000s and is currently under consideration as a formal addition to the geologic record (Crutzen and Stoermer, 2000; Lewis and Maslin, 2015; Steffen et al., 2015), the term has been increasingly taken up in the social sciences (Taylor, 2017), humanities (Braidotti, 2013; Morton, 2016), arts (Turpin and Davis, 2015), and the public domain as a subsuming descriptor for the current age. The sociomaterial impacts of the Anthropocene in the lives of children and young people has also become a focus of environmental education and childhood studies in recent years, often drawing on posthumanist and new materialist theoretical approaches that open inquiry to more-than-human ecologies and relationalities (c.f. Taylor and Pacini-Ketchabaw, 2015; Malone, 2016a; Rousell et al., 2017; Somerville, 2017).

Despite the rapid uptake of the Anthropocene concept, many scholars have hesitated to adopt a term so saturated with "Anthropos" and its aftertastes of human (white, male) dominance, supremacy, and exceptionalism (Colebrook, 2014; Demos, 2017). Alternative figurations of the current age have emerged, variously emphasising the total subsumption of Earthly processes under a capitalist political economy in terms of "Capitalocene" (Moore, 2017); the chthonic, nonhuman powers of the Earth itself as an emergence of the "Chthulucene" (Haraway, 2016); and the ubiquitous technical mediation of Earth processes in terms of a "technocene" (Hörl, 2017), or "datacene" (Rousell, 2020).

Recent decolonial scholarship further critiques the Anthropocene as a conceptual and material artefact of Western (mis)thought, which erases colonial histories of violence and subjection through the figuration of a universal humanity equally responsible for ecological catastrophe (Horton, 2017; Todd, 2015; Yusoff, 2018). Climate change can be understood as the direct result of a capitalistic-colonial logic of exploitation and manipulation of life according to the predatory values of individual monetisation and profit (Guattari, 2008; Rolnik, 2017; Yusoff, 2018). This logic operates according to a market-driven extraction and consumption of geological, biological, and sociogenic matter (Wynter and McKittrick, 2015), extracting not only the chemical materials and "flesh" needed to sustain unregulated economic growth, but also, and by the same logic, harvesting the energy, creativity, feeling, and intensity that sustains human and nonhuman life. Rolnik (2017) describes this logic of extraction as "the pimping of life", which operates through a perverse abuse and commodification of life's vitality as a creative force.

> What distinguishes the colonial-capitalistic system is the pimping of life as a force for creation and transmutation.... The vital force of the entire biosphere is expropriated and corrupted by that system: the land, the air, the water, the sky, the plants, the animals, and the human species ... the effect is the dis-empowerment of life, interrupting its germinal process, destroying the vital energy sources of the biosphere – which, in humans, includes the subjective resources for its preservation.
>
> *(Rolnik, 2017, pp. 4–5)*

As a planetary-scale symptom of this capitalistic and colonial logic of extraction, climate change is now kicking back with a violent force that simultaneously destabilises and reinforces the inherent injustice of colonial regimes that fuelled its emergence. While the Anthropocene disrupts anthropocentric claims of human separation and mastery over nature, its most severe effects are currently impacting communities which have historically been subjected to capitalistic-colonial violence, including Indigenous, Black, Minority Ethnic, non-western, and migrant communities (Baldwin and Bettini, 2017), as well as children and young people worldwide (Rousell et al., 2020).

In her book *A Billion Black Anthropocenes or None*, Yusoff (2018) argues that the geologic extraction and displacement of mineral and oil deposits is indissociably linked to the extraction and displacement of human flesh as "fungible" matter under the conditions of chattel slavery.

> As the Anthropocene proclaims the language of species life – *anthropos*-through a universalist geologic commons, it neatly erases histories of racism that were incubated through the regulatory structure of geologic relations ... The human and its subcategory, the inhuman, are historically relational to a discourse of setter-colonial rights and the material practices of extraction, which is to say that the categorisation of matter is a spatial execution, of place, land, and person cut from relation through geographic displacement (and relocation through forced settlement and transatlantic slavery).
>
> *(Yusoff, 2018, p. 2)*

Yusof refers to the earliest proposed geologic marker for the Anthropocene in 1610, the point at which the mass European invasion of North and South America reached geologic magnitude in the stratigraphic record. Research by Lewis and Maslin (2015) suggests that the colonisation of the Americas beginning in 1492 led to an "irreversible cross-ocean movement of species", an intermixing of biota from continents separated by previously uncrossable bodies of water, generating "one of the very few human-induced changes that is likely captured today in stratigraphic records and is a near-permanent change to Earth" (p. 144). It is further claimed that the ecological destablisations wrought by colonisation and the transatlantic slave trade permanently "set the Earth system on a new trajectory" (p. 144) that will be visible in the planet's stratigraphic layers for millions of years.

By 1610 the colonisation of the Americas and the global slave trade had been underway for 150 years, and the Indigenous population of the Americas had been decimated from an estimated 56 million in 1492 to only 6 million in 1610 (Lewis and Maslin, 2015). This mass genocide and displacement of Indigenous peoples is preserved in the geologic carbon record as the first verifiable "signal" of irreversible human impact on the trajectory of the Earth system as a whole. As Yusoff writes,

> this spike of brutality, sadism, and death, coupled with the subsequent dispossession of indigenous peoples from their land and the beginnings of industrial global slavery, enacts a foundational spatial inscription of colonialism (and race) into a monument of global environmental change.
>
> *(Yusoff, 2018, p. 32)*

Rather than indexing the inadvertent results of humanity's global expansion and enterprise, the "golden spike" in the geologic record that scientists attribute to the onset of the Anthropocene bears the mark of calculated and sustained pathological violence, a necropolitical project of planetary subjection that continues to feed on the racialised extraction of human and nonhuman life (Mbembe, 2019).

The urgency of responding to what has been named "Anthropocene" pushes us to think beyond the material-discursive contortions of climate change to a fundamental crisis in subjectivity and ethics, a crisis in how to live and imagine a life beyond the colonial-capitalistic subject (Grosz, 2017; Rolnik, 2017). This involves an acknowledgement that the vision of the Anthropocene as a new geologic age of humanist planetary stewardship is deeply flawed, resting on a false futurity premised on the erasure of an unthinkably brutal colonial history and its afterlives in the present (Hartman, 1997; Todd, 2015). Rousell et al., (2020, p. 4) argue that the Anthropocene has become "an empty signifier for an unnameable epoch in which the colonial-capitalistic regime achieves planetary dominance, while attempting to erase its own (violent, barbaric) tracks with the promise of a 'humanity' in common". They offer examples of how children encountered the Anthropocene as a "sense-event" in a university museum, and developed a series of artistic interventions aimed at decolonising the unconscious thresholds of museological claims to mastery, neutrality, and universal knowledge (see also de Freitas et al., 2022; Hohti et al., 2021). This book pursues a similar trajectory with regard to the Anthropocene, which serves not as a universal descriptor for an age of human dominance, but rather as a zone of critical and creative entanglement. As Trafí-Prats (2020, p. 11) writes, attending to the tensions and complexities of decolonisation in the Anthropocene "is neither about finding a home nor about proposing a fix, but about sensing vulnerability, fragility, incommensurability, and the struggle for articulation". To encounter the Anthropocene is to encounter planetary scale trauma, violence, fear, uncertainty, instability, extinction, and death. And yet, our encounters of the Anthropocene with children and young people have also cultivated a shared sense of wonder, vulnerability, openness, and sensitivity to worlds of experience that have been systemically silenced, exploited, and excluded.

Posthuman Childhood(s)

> He was only a child in the eyes of the public. Not likely to be taken seriously.
>
> **Jasmyne, age 12 years, CC+Me Researcher**

Scholars working the intersections between the fields of childhood studies and environmental education have proposed that the onset of climate change calls for new conceptualisations of childhood and youth (Malone, 2016b; Murris, 2016; Somerville, 2017; Taylor, 2017), along with new research methodologies and techniques attuned to the complex dynamics of children's social and ecological worlds (Cutter-Mackenzie-Knowles and Rousell, 2020). As the relentless acceleration of social, technological, and environmental change continues to erode the humanist foundations of dualistic Western thought and its systemic structures, we are now witnessing a proliferation of biosocial (Lee, 2013; Youdell, 2017), new materialist (Hultman and Lenz Taguchi, 2010; Malone, 2016a; Murris, 2016; Sheldon, 2016), decolonial (Nxumalo and Cedillo, 2017), socioecological (Brown et al., 2014) and post-developmental (Blaise, 2014) framings of childhood which could be broadly described as "posthumanist". In many cases this work has involved the reinvigoration of a more-than-human political ecology or commons, as an open assemblage of human and nonhuman agencies that comprise the messy and entangled lives of children and young people (Taylor, 2013). Posthumanist reconceptualisations of childhood have foregrounded the pluralistic and highly situated nature(s) of childhood(s), showing how childhoods are lived and conditioned differently within particular assemblages of place, race, class, gender, cultural, and sexual difference (McCoy et al., 2018; Nxumalo and Cedillo, 2017; Trafí-Prats, 2020).

This book contributes to posthumanist reconceptualisations of childhood and youth through the cultivation of alternative aesthetic figurations and sociopolitical imaginaries of climate change emerging from collaborations with children and young people as researchers. As Rebekah Sheldon (2016, n.p.) asks in her book *The Child to Come*, "why, when we reach out to grasp the future of the planet, do we find ourselves instead clutching the child?" Sheldon argues that figurations of childhood and youth in contemporary western societies are intricately bound up with aspirations for a "future humanity" expected to redeem the unethical transgressions, pathologies, and climatological failures of a colonial-capitalistic past and present. Casting the weight of planetary stewardship and ecological responsibility onto the child effectively performs an "incision of the planet into the child-relationship", such that "the figuration of the child as the self-similar issue of the present, the safe space of human prosperity and a return to manageable nature, forecloses the mutational and the reproductive" (Sheldon, 2016, n.p.).

While children and young people are among the least responsible for perpetuating the current climate crisis, they are consistently positioned as future "resources" in whom the public invests to secure the ongoing viability of human life according to particular cultural and political regimes (Lee, 2013). Children remain second-class (inhuman) citizens without political voice or agency until they have received sufficient "humanisation" under the auspices of K-12 schooling, at which point they are considered to be rational beings capable of participating in political life (Snaza, 2015). The transcendent aim of Western education is the humanisation of the child through the social ordering of experience into the figure of a rational subject (Rolnik, 2017; Snaza and Weaver, 2015). The contemporary figuration of the child is therefore gridded between the neoliberal investment in childhood as "speculative commodity future" (Lee, 2013; Massumi, 2018), and the protection of childhood innocence against perpetration from "outside" a particular ontological regime of human subjectivity and social ordering (Murris, 2016; Sheldon, 2016; Taylor, 2013). This gridlocked figuration of childhood and youth in contemporary Western societies effectively works to prevent any kind of systemic, qualitative transformation by prohibiting the social reproduction of alternative sensibilities, ideas, and actions that deviate from normative standards of the human subject as a self-enclosed individual.

This has been particularly evident in the backlash against the 2019 youth school strikes resisting climate inaction in Australia. School leaders across Australia banded together to publicly condemn the school strikes, specifically stating that young people were "not ready" to be exposed to the complex and "debatable" issue of climate change, and that they should instead focus on their academic and sporting achievement at school (Boulianne et al., 2020). Many schools banned students from wearing pins, badges, or any other indication of their support for the climate protests at school, and penalised any students who actively participated in the movement. Implicit in these statements and actions is the determination that children and young people are simply incapable of thoughtful engagement with climate change, that they are politically innocent, vulnerable, irrational, and naïve, and that their political status is restricted to that of "educables", capable only of being educated by their betters.

Despite these institutional attempts to quash and control children and young people's political agency, over 1.4 million children and young people walked out of school to protest against climate inaction on March 15, 2019. These strikes were organised entirely by children and young people through social media and other web-based platforms. The success of this movement has created a new public platform not only for youth voice and political agency, but also for the cultivation of alternative ethical and political framings, surfacing new opportunities for young people to engage in political thought that situates human life and flourishing as but one element within a more-than-human ecological commons. Refuting public perceptions of youth as politically disaffected and disengaged, the strikes demonstrate young people's capacity to self-organise dissident political action on a global scale, using one of the few political weapons at their disposal: *refusal*. By refusing to attend school and "be educated", often in direct conflict with local authorities, young people have demonstrated a refusal to participate in a political and educational system which they perceive as inherently unjust, intolerable, and with respect to climate change, cataclysmic.

The intractable problematic of climate change is arguably the most salient example of how humanist imaginaries of progressive social improvement and mastery over nature have failed to contain a vastly more distributed and indeterminate reality of ecological entanglement. This book seeks to seed a series of speculative alternatives to this failed investment in humanist educational futures. The question remains as to how the recent upsurge of youth activism movements might intersect with climate change education and its associated research. How might climate change education in schools and universities engage with this new youth politics? What are the potentials of posthumanist modes of climate change education and research, including those that work directly with children and young people in the cultivation of new political and aesthetic imaginaries?

The Minor Sphere of Micropolitical Insurrection

In taking an oppositional stance against institutional authority, mass movements such as #ClimateStrike4Action demonstrate the viral power of political affects, images, and ideas in contemporary economies of attention regulated largely by digital and social media (Rousell et al., 2021). While the recent mass climate strikes have brought unprecedented public attention to young people's political agency with respect to climate change, such macropolitical movements tend to perpetuate a liberal, humanist image of the bounded subject as operationalised through oppositional models of collectivity as an undifferentiated mass or aggregate (us vs them). In its oppositional form, a macropolitical movement is defined by its opposition to an established social order, and thus overshadows the opportunity to seed alter-political *alternatives* for living and experiencing the world beyond the bounded interiority of the humanist subject (Massumi, 2018).

We can see this, for instance, in the continuous reterritorialisation of the #ClimateStrike4Action movement onto the identitarian image of Greta Thunberg as the movement's leader. This demonstrates how mass political movements can often obscure the potentials for less visible, micropolitical forms of collective resistance and transformation from *within* institutional regimes and intensive ecologies of experience. There is a tendency, in this sense, for mass environmental movements to reify the very politics of individualism which they seek to challenge.

In his essay *The Three Ecologies*, Guattari (2008) describes the need for micropolitical movements that transform the very conditions for individual and collective subjectivity from within public institutions associated with education, health, the arts, culture, architecture, municipal government, and social services. He describes this as a process of both "resingularisation" and "heterogenesis", involving transversal practices that engender the micropolitical transformation of subjectivity across the tri-ecological realms of the environment, the socius, and the conceptual. For Guattari, the three ecologies operate through an eco-logic of intensity, rather than a logic of discursive categories that presuppose a totalising regime of individual subject-identities.

> While the logic of discursive sets endeavours to completely delimit its objects, the logic of intensities, or eco-logic, is concerned only with the movement

> and intensity of evolutive processes. Process, which I oppose here to system or structure, strives to capture existence in the very act of its constitution, definition, and deterritorialisation.
>
> *(Guattari, 2008, p. 30)*

Brazilian psychoanalyst and art theorist Suely Rolnik (2017) further delineates the differences between macropolitical, discursive movements and what she terms the "sphere of micropolitical insurrection", which operates through a logic of intensity. Rolnik argues that macropolitical movements hinge on an image of the subject as bounded and separated from the vitality and immanence of life. Macropolitics positions the subject as subordinate to the power relations that order the social world of human affairs, and aims to redistribute such power relations in ways that are perceived to be more equitable, humane, and empowering. The sphere of the macropolitical can thus only be conceived and enacted by humans because it is founded on the identity position of the humanist subject, "causing us to remain captive to the logic of the very system we seek to combat" (p. 8).

Only micropolitics, according to Rolnik (2017), is capable of resisting the capitalistic and colonial abuse of life by mobilising both human and non-human subjectivities, indeed, "all the elements of the biosphere that rebel against violence toward life" (p. 8). Rolnik describes how non-human processes immediately recognise and combat the abuse of life, including a river in Brazil that went underground to evade toxic contamination on the surface, or trees that alter their flowering cycles to prevent sterility from pollution. These examples show how the micropolitical sphere of insurrection is orientated towards multi-species flourishing, as a more-than-human response and rebellion against capitalistic and colonial regimes that abuse the vitality of Earthly life. Rolnik describes how micropolitical insurrection operates through a "potentialisation of life … by affirming life in its generative essence", rather than through an "empowerment" of the bounded subject which only attempts to redistribute power relations among fixed determinations of identity (p. 9). For Rolnik, micropolitics involves a complex and subtle process of "decolonising the unconscious" by gradually dissolving the grip of the capitalistic and colonial regime both "within us and outside of us" (p. 10). These include micropolitical practices that cultivate our collective vulnerability and ecological sensitivity, activating affective modes of bodily knowing and creative experimentation by "reimagining the world in each gesture, each word, each relation, each mode of existence" (p. 10).

We discern in Rolnik's account that the macropolitical is not necessarily larger in scale or significance than the micropolitical. The macropolitical is just more *visible*, while the micropolitical tends towards imperceptibility. A mass climate protest, for instance, often takes shape as a macropolitical aggregate of subjects that is highly visible. However, the micropolitical relations and affective bonds between each of those protestors and the entire biosphere are largely invisible, but not necessarily smaller in scale, less powerful, or less significant. As anonymous anarchist collective the Invisible Committee writes, even the smallest act of insurrection connects us to the planetary epoch in which we find ourselves.

> However localized it may be, every insurrection gestures beyond itself; it contains something global from the outset. It raises us together to the level of the epoch. But the epoch is also what we find deep within us, that is, when we're willing to descend that far, when we immerse ourselves in what we're experiencing, seeing, feeling, perceiving.... The epoch must be sought deep within each situation and deep within each person.
>
> *(The Invisible Committee, 2014, pp. 5–6)*

Rolnik's vision of micropolitical insurrection also resonates with the conceptual figure of the "minor", as introduced by Deleuze and Guattari (1986) in reference to minor forms of literature. Where majoritarian or State regimes present a "constant and homogenous system", a minoritarian politics operates through "subsystems ... seeds [and] crystals of becoming whose value is to trigger uncontrollable movements and deterritorialisations of the mean or majority" (Deleuze and Guattari, 1986, p. 106). As Manning (2016, p. 1) elaborates in the early pages of her book *The Minor Gesture*:

> The major is a structural tendency that organises itself according to predetermined definitions of value. The minor is a force that courses through it, unmooring its structural integrity, problematising its normative standards.... The minor is a continual variation on experience. It has a mobility not given to the major: its rhythms are not controlled by a pre-existing structure, but are open to flux.
>
> *(Manning, 2016, p. 1)*

The minor gesture of seeding and triggering micropolitical shifts in feeling and thought is pivotal for our work in this book, and how we understand our work with children and young people in terms of an ecological aesthetics. We are interested in thinking and writing *with* the creative practices of children as micropolitical actions that contribute to our shared inquiry, rather than writing *about* children as bounded "subjects" or "objects" of social research. By striving to think and work in the sphere of the aesthetic and the micropolitical, we aim to show how the minor gestures of children and young people's critical and creative research can proliferate virally through majoritarian systems of educational governance, gradually decolonising calcified regimes of subjectification, compliance, and business-as-usual managerialism by carving out new trajectories for alternative futures.

Situating CC+Me

Our engagement with the aesthetic and micropolitical dimensions of young people's encounters with climate change is necessarily situated in complex relations with particular places, cultures, and milieus, both human and more than human. Before we explore the specific theories, practices, and events that constituted the research in more detail throughout the book, it is important to briefly contextualise the Indigenous and non-Indigenous heritage of the lands on which the project was conducted. The Northern region of NSW in which the project took place was originally inhabited by the Indigenous peoples of the Bundjalung nation, a large conglomeration of over 15 Aboriginal clans who historically occupied the coastal region stretching from the Clarence river in NSW up to the Beaudesert region of what is now South-east Queensland. The Bundjalung nation incorporates an extraordinary plurality of cultural, language, and spiritual practices and traditions which co-existed in the region for over 12,000 years.

Bundjalung Country was claimed by James Cook when his boat passed the region in 1770, with the colonial invasion of the region reaching critical mass in the mid-1800s with the pursuit of valuable red cedar trees and the claiming of farmland. The invasion resulted in mass deforestations of the coastal rainforest and the decimation of Indigenous peoples and ways of life. The history of the area is marked by a series of large-scale massacres of Indigenous people by European invaders in the mid-1800s and early 1900s, and there is evidence of attempts to erase these events from the public historical record.

Indigenous people now make up approximately 1.6% of the region's population, although recent census documents suggest this percentage may be growing. The Northern NSW region is known for its physical beauty and is a coastal tourist destination which also serves as the region's primary industry. Several World Heritage national parks are located in the region which include marine and rainforest eco-systems with species, communities, and relations of life found nowhere else in the world. The region is situated in a subtropical latitude with typically long, hot summers and short, mild winters. The socio-economic demographics are largely divided across Shires of greater or lesser distance to the coast. Coastal areas such as Byron Shire attract the highest median housing prices anywhere in Australia, while rural areas such as Kyogle Shire offer much cheaper lifestyles. There is a sense of seasonal itinerancy that accompanies the shifting patterns of weather in tandem with the ebb and flow of tourist seasons. But irrespective of the swell of tourists that quadruples the population of the region every summer, the vast majority of the region's demographic is of white Australian or white Anglo-European decent.

We offer this brief contextual overview in order to highlight some of the ways that living, studying, and working in the Northern Rivers region of NSW is a privilege that is largely determined by racialised and class-based conditions. The schools that participated in the CC+Me project were located across the coastal Shires of Tweed and Ballina, and the inland Shires of Lismore and Kyogle. There were notable differences between the socio-economic conditions of these schools, as well as cultural differences between the suburban coast and the rural villages 100 miles inland. Despite these geographical differences, we were struck by the overwhelming whiteness of the schools that we visited across these Shires. While our workshops were open to all students between the ages of 9 and 14 attending the six schools we worked with, only 3 of the 135 children and young people who participated in the project identified as being Indigenous.

We mention this not in an attempt to reinforce markers of race and identity on the CC+Me researchers but, rather, to acknowledge that elements of racialisation and socio-economic precarity are part of the ontogenic and sociogenic milieus (Wynter and McKittrick, 2015) of the region, and, by extension, the CC+Me project. The environmental beauty of the region can inspire and enable alternative lifestyles and creative practices, but that same beauty can also be disorientating, even blinding. Qualities of life and access to rich coastal and rainforest environments can too easily be experienced as something simply *deserved*, masking the reality that these environments have been violently *taken*, and are continuously *repossessed* by regimes that unevenly distribute material access, power, and resources along geosocial (Yusoff, 2018) and biopolitical (Weheliye, 2014) lines of racialisation and socio-economic status.

As researchers of mixed European, white American, Jewish, and white Australian ancestry, we acknowledge our own complicity in the privilege of white access and inhabitation of this region, and that the violent colonial history of occupation forms part of the very condition of possibility for the CC+Me project. We also maintain that the CC+Me project has a social and affective life that extends and proliferates beyond the locality of its inception, as it continues to proliferate along alternative trajectories of collective thought, feeling, and sociality. Evoking these situated conditions of inquiry helps to resist a majoritarian claim to a universal "we", which might attempt to generalise the realities of children and young people, or even "humanity", as a neutral or universal signifier. Our use of the plural pronoun throughout this book aims instead to invoke a minor "we" articulated through a distinctive collective of two adult researchers and 135 young researchers who worked together over three years to generate new climate child imaginaries through aesthetic practices of co-existence. To this end, the book is not written as a representation of children and young people's (or our own) identities, but, rather, as a shared process of encountering the intractable problematic of the Anthropocene *together*, through the practices of a diversely situated, intergenerational research collective.

Orientating Towards This Book

This book is organised around seven chapters offering alternative imaginings of climate change education as a field that can foster and achieve a new synthesis of creative, critical, philosophical, and empirical practices and orientations. The book connects climate change education with posthumanist studies of childhood in the social sciences and environmental humanities, while also offering distinctive opportunities for readers to encounter new theoretical and methodological approaches for collaborative art, inquiry, and learning with children. This opening chapter has offered an initial orientation to our approach through a critical posthumanist engagement with the concepts of climate change imaginaries, the Anthropocene, and micropolitics. We have highlighted our specific engagements with fields such as posthumanism, process philosophy, affect theory, and decolonial studies as key theoretical touchstones throughout the book. We also introduced and briefly situated the Climate Change and Me project that forms the empirical basis for the book.

In Chapter 2 we critically review existing literature on climate change education, and identify the need for posthumanist approaches that enable children and young people to engage with the issue affectively, critically, and creatively. We critically consider the moralisation and politicisation of climate change in children's everyday lives, where matters of climate fact, value, and concern become inextricably intertwined. Drawing on multiple genealogies of work in science and technology studies, eco-feminism, and the environmental posthumanities, we situate our approach to climate change education as a project of learning to live with care in times of radical unsustainability and precarity across multiple scales of co-existence.

Chapter 3 focuses on developing the methodological concept of the "research playspace". Drawing on environmental theories of play, affect, and post-developmental childhood studies, we describe how the research playspace was constructed at various stages of the project. Winnicott's concepts of transitional space and transitional objects are used to develop an ecological understanding of how children came to participate in the project as researchers. We also consider the role of creative and affective pedagogies in shaping our methodological approach, where teaching and learning are decoupled from individual identities and reconfigured as environmentally distributed processes. This leads to a consideration of children's use of digital video as a transitional medium for elaborating complex insights into climate child imaginaries, and detailed accounts of how our emergent methods led to exhibitions and curriculum resources co-produced with children.

Having outlined the context and workings of the Climate Change and Me project in earlier chapters, Chapter 4 turns to Whitehead's philosophy of nature to develop a speculative understanding of children and young people's creative practices through their participation in more-than-human ecologies of experience. Here we begin the task of developing an *ecological aesthetics of childhood* which connects the intensive movements of childhood experience to the "open whole" of nature. This includes a discussion of Whitehead's key concepts of becoming, prehension, and feeling in conversation with current work on aesthetics at the intersections of new materialisms, life sciences, and media studies. We develop the distinction between *uncommon* and *common* worlds as an analytic for grasping children's dynamic relationships with/in nature as a "total environment", with a focus on children's environmental photographs and poetry as examples of creaturely aesthetic attunement that resist the bifurcation of nature into subjective and objective domains.

Chapter 5 continues to build on the theoretical engagement with Whitehead's philosophy in Chapter 4 to elaborate the affective power of images in the construction of climate child imaginaries. This entails a retheorisation of the image as what Whitehead terms a "vector of feeling" which feeds affective intensity from the past into the future. Drawing on examples of children's image-making practices that shift across digital and analogue formats, we develop analytic methods for reading images as surfaces of experience rather than representations. This leads to a reconsideration of the role of perception in the construction of climate child imaginaries, where perception is taken as an affective process of constructive feeling that, at the limit, opens onto the vast material-causal infrastructures of nature.

In Chapter 6 we shift our analytic focus from children's image production to the study of children's speculative writing practices. Noting the recent turn to speculative fiction as a "new empirical" research method in education and the social sciences more broadly, we open a discussion on the role of speculative character- and world-building in the construction of alternative climate child imaginaries. Our discussion draws specifically on the emergence of a collaborative online fiction experiment led by one of the child researchers in the Climate Change and Me project. We describe how fictional characters from the future began to appear in entries on the project website, and developed into an intricate speculative world that evolved in parallel with our project's development in real time. We turn to Deleuze and Guattari's geophilosophy and Erin Manning's concept of the "minor gesture" in order to develop an analytic for grasping this irruption of speculative fabulation from the middle of our project. Through a close reading of the aberrant characterisations and often violent scenes that children introduced into the project, we follow their elaboration of a dark ecological aesthetic for living on an occulted and decaying planet, where the radical transmutation of bodily and affective life remains one of the only possibilities for resisting and surviving climate change.

Chapter 7 then concludes the book with a series of generative openings and propositions for future research that have arisen in the years since the Climate Change and Me project concluded. We describe generative openings into the digital through the development of a climate change education web App in collaboration with young people across Australia. We also offer some examples of new international collaborations between children, artists, and researchers responding to climate change in cities through interventions in galleries and museums. We then close with a consideration of generative openings through Indigenous concepts of Country (climate Country), and opportunities for respectful alliances between Indigenous and non-Indigenous communities in developing climate responsive practices and imaginaries for a posthuman future.

References

Ahmed, S. (2008). Sociable happiness. *Emotion, Space and Society*, *1*(1), 10–13.

Appadurai, A. (1996). *Modernity at large: Cultural dimensions of globalization*. University of Minnesota Press.

Atkinson, D. (2017). *Art, disobedience, and ethics: The adventure of pedagogy*. Springer.

Baldwin, A., & Bettini, G. (2017). *Life adrift: Climate change, migration, critique*. Rowman & Littlefield.

Berlant, L. (2011). *Cruel optimism*. Duke University Press.

Bertelson, L., & Murphie, A. (2010). An ethics of everyday infinities: Felix Guattari on affect and the refrain. In M. Gregg, & G. J. Seigworth (Eds.), *The affect theory reader*, (pp. 138–160). Duke University Press.

Bignall, S., & Braidotti, R. (2019). Posthuman systems. In R. Braidotti, & S. Bignall (Eds.), *Posthuman ecologies: Complexity and process after Deleuze* (pp. 1–22). Rowman & Littlefield.

Blaise, M. (2014). Interfering with gendered development: A timely intervention. *International Journal of Early Childhood, 46*(3), 317–326. https://doi.org/10.1007/s13158-014-0122-9.

Boulianne, S., Lalancette, M., & Ilkiw, D. (2020). "School strike 4 climate": Social media and the international youth protest on climate change. *Media and Communication, 8*(2), 208–218.

Braidotti, R. (2013). *The posthuman*. Polity Press.

Brown, T., Jeanes, R., & Cutter-Mackenzie, A. (2014). Social ecology as education. In B. Watchow, R. Jeanes, L. Alfrey, T. Brown, A. Cutter-Mackenzie, & J. O'Connor (Eds.), *The socioecological educator* (pp. 23–45). Springer. https://doi.org/10.1007/978-94-007-7167-3.

Cajete, G. (2006). Western science and the loss of natural creativity. In D. T. Jacobs (Ed.), *Unlearning the language of conquest* (pp. 247–259). University of Texas Press.

Castoriadis, C. (1987). *The imaginary institution of society* (K. Blamey, Trans). MIT Press.

Cole, D. R., & Mirzaei Rafe, M. (2017). Conceptual ecologies for educational research through Deleuze, Guattari and Whitehead. *International Journal of Qualitative Studies in Education, 30*(9), 849–862.

Cole, D. R., & Somerville, M. (2017). Thinking school curriculum through country with Deleuze and Whitehead: A process-based synthesis. In C. Naughton, G. Biesta, & D. Cole (Eds.), *Art, artists and pedagogy* (pp. 71–82). Routledge.

Colebrook, C. (2014). *Death of the PostHuman: Essays on extinction* (Vol. 1). Open Humanities Press.

Crutzen, P. J., & Stoermer, E. F. (2000). Global change newsletter. *The Anthropocene, 41*, 17–18.

Cutter-Mackenzie-Knowles, A., & Rousell, D. (2020). The mesh of playing, theorising and researching in the reality of climate change: Children's childhoodnature research playspaces. In A. Cutter-Mackenzie-Knowles, K. Malone, & E. Barratt Hacking (Eds.), *Research handbook on childhoodnature: Assemblages of childhood and nature research* (pp. 199–222). Springer Nature.

Deleuze, G., & Guattari, F. (1986). *Kafka: Toward a minor literature* (D. Polan, Trans). University of Minnesota Press.

Deleuze, G., & Guattari, F. (1987). *A thousand plateaus* (B. Massumi, Trans.). University of Minnesota Press.

Demos, T. J. (2017). *Against the anthropocene. Visual culture and environment today* (p. 132). Sternberg Press.

Demos, T. J. (2020). *Beyond the world's end: Arts of living at the crossing*. Duke University Press.

Ellsworth, E., & Kruse, J. (2012). Introduction. In E. Ellsworth, & J. Kruse (Eds.), *Making the geologic now: Responses to material conditions of contemporary life* (pp. 5–26). Punctum Books.

de Freitas, E., Trafí-Prats, L., Rousell, D., & Hohti, R. (2022). A poetics of opacity: Towards a new ethics of participation in gallery-based art projects with young people. In L. Trafí-Prats, & A. Castro-Varella (Eds.), *Visual participatory arts based research in the city* (pp. 126–142). Routledge.

FridaysForFuture (2019). https://fridaysforfuture.org/ Accessed January, 2020.

Gabrys, J. (2016). *Program earth: Environmental sensing technology and the making of a computational planet*. University of Minnesota Press.

Gatens, M., & Lloyd, G. (1999). *Collective imaginings: Spinoza, past & present*. Routledge.

Grosz, E. (2017). *The incorporeal: Ontology, ethics, and the limits of materialism*. Columbia University Press.

Guattari, F. (2008). *The three ecologies*. Continuum.

Haraway, D. (2016). *Staying with the trouble: Making kin in the Chthulucene*. Duke University Press.
Hartman, S. V. (1997). *Scenes of subjection: Terror, slavery, and self-making in nineteenth-century America*. Oxford University Press on Demand.
Hickey-Moody, A. (2013). *Youth, arts, and education: Reassembling subjectivity through affect*. Routledge.
Hickey-Moody, A., Harwood, V., & McMahon, S. (2016). Feeling futures: The embodied imagination and intensive time. In D. Bland (Ed.), *Imagination for inclusion: Diverse contexts of educational practice* (pp. 128–140). Routledge.
Hohti, R., Rousell, D., MacLure, M., & Chalk, H. L. (2021). Atmospheres of the Anthropocene. Sensing and rerouting dis/inheritances in a university museum with young people. *Children's Geographies*, 1–14.
Hörl, E. (2017). Introduction to general ecology. In E. Hörl, & J. Burton (Eds.), *General ecology: The new ecological paradigm* (pp. 1–73). Bloomsbury.
Horton, J. L. (2017). Indigenous artists against the anthropocene. *Art Journal*, *76*, 48–69. https://doi.org/10.1080/00043249.2017.1367192.
Hultman, K., & Lenz Taguchi, H. (2010). Challenging anthropocentric analysis of visual data: A relational materialist methodological approach to educational research. *International Journal of Qualitative Studies in Education*, *23*(5), 525–542.
Lee, N. (2013). *Childhood and biopolitics: Climate change, life processes and human futures*. Palgrave Macmillan.
Lewis, S. L., & Maslin, M. A. (2015). A transparent framework for defining the Anthropocene Epoch. *The Anthropocene Review*, *2*(2), 128–146.
Malone, K. (2016a). Reconsidering children's encounters with nature and place using posthumanism. *Australian Journal of Environmental Education*, *32*(1), 42–56.
Malone, K. (2016b). Theorizing a child–dog encounter in the slums of La Paz using post-humanistic approaches in order to disrupt universalisms in current 'child in nature' debates. *Children's Geographies*, *14*(4), 390–407. https://doi.org/10.1080/14733285.2015.1077369.
Manning, E. (2016). *The minor gesture*. Duke University Press.
Manning, E., & Massumi, B. (2014). *Thought in the act: Passages in the ecology of experience*. University of Minnesota Press.
Massumi, B. (2011). *Semblance and event: Activist philosophy and the occurrent arts*. MIT Press.
Massumi, B. (2018). *99 Theses on the revaluation of value: A postcapitalist manifesto*. University of Minnesota Press.
Mazzei, L. A. (2021). Speculative inquiry: Thinking with Whitehead. *Qualitative Inquiry*, *27*(5), 554–566.
Mbembe, A. (2019). *Necropolitics*. Duke University Press.
McCoy, K., Tuck, L., & McKenzie, M. (2018). *Rethinking pedagogies of place from Indigenous, postcolonial, and decolonizing perspectives*. Routledge.
McNeill, J. R., & Engelke, P. (2014). *The great acceleration: An environmental history of the Anthropocene since 1945*. Harvard University Press.
McNeill, K. L., & M. H. Vaughn. (2012). Urban high school students' critical science agency: Conceptual understandings and environmental actions around climate change. *Research in Science Education*, *42*(2), 373–399.
Moore, J. W. (2017). The capitalocene, part I: On the nature and origins of our ecological crisis. *The Journal of peasant studies*, *44*(3), 594–630.
Morton, T. (2016). *Dark ecology: For a logic of future coexistence*. Columbia University Press.
Murris, K. (2016). *The posthuman child: Educational transformation through philosophy with picturebooks*. Routledge.

Nxumalo, F., & Cedillo, S. (2017). Decolonizing place in early childhood studies: Thinking with Indigenous onto-epistemologies and Black feminist geographies. *Global Studies of Childhood*, 7(2), 99–112.

O'Brien, K., Selboe, E., & Hayward, B. M. (2018). Exploring youth activism on climate change. *Ecology and Society*, *23*(3).

Papoulias, C., & Callard, F. (2010). Biology's gift: Interrogating the turn to affect. *Body & Society*, *16*(1), 29–56.

Ritchie, J. (2021). Movement from the margins to global recognition: Climate change activism by young people and in particular indigenous youth. *International Studies in Sociology of Education*, *30*(1–2), 53–72.

Rolnik, S. (2017). The spheres of insurrection: Suggestions for combating the pimping of life. *E-flux Journal*, *86*, 1–15.

Rousell, D. (2020). A map you can walk into: Immersive cartography and the speculative potentials of data. *Qualitative inquiry*. https://doi.org/10.1177/1077800420935927.

Rousell, D. (2021). *Immersive cartography and post-qualitative inquiry: A speculative adventure in research-creation*. Routledge.

Rousell, D., & Cutter-Mackenzie-Knowles, A. (2020). A systematic review of climate change education: Giving children and young people a 'voice' and a 'hand' in redressing climate change. *Children's Geographies*, *18*(2), 191–208. https://doi.org/10.1080/14733285.2019.1614532.

Rousell, D., Cutter-Mackenzie-Knowles, A., & Foster, J. (2017). Children of an Earth to come: Speculative fiction, geophilosophy and climate change education research. *Educational Studies*, *53*(6), 654–669. https://doi.org/10.1080/00131946.2017.1369086.

Rousell, D., Hohti, R., MacLure, M., & Chalk, H. L. (2020). Blots on the anthropocene: Micropolitical interventions with young people in a university museum. *Cultural Studies ↔ Critical Methodologies*, *21*, 27–40.

Rousell, D., Ryan, E., Bauer-Nilsen, B., & Lai, R. (2022). Animist pedagogies and the endings of worlds: Rituals for the pluriverse. In P. Burnard, L. MacKinlay, D. Rousell, & T. Dragovic (Eds.) *Doing rebellious research in and beyond the academy* (pp. 263–281). Brill.

Rousell, D., Wijesinghe, T., Cutter-Mackenzie-Knowles, A., & Osborn, M. (2021). Digital media, political affect, and a youth to come: Rethinking climate change education through Deleuzian dramatisation. *Educational Review*. https://doi.org/10.1080/00131911.2021.1965959.

Seigworth, G. J., & Gregg, M. (2010). An inventory of shimmers. In M. Gregg, & G. J. Seigworth (Eds.), *The affect theory reader*. (pp. 1–28). Duke University Press.

Shaviro, S. (2015). *Discognition*. Repeater Books.

Sheldon, R. (2016). *The child to come: Life after the human catastrophe*. University of Minnesota Press.

Shepardson, D. P., D. Niyogi, A. Roychoudhury, & A. Hirsch. (2012). Conceptualizing climate change in the context of a climate system: Implications for climate change and environmental education. *Environmental Education Research*, *18*(3), 323–352.

Snaza, N. (2015). Toward a genealogy of educational humanism. In N. Snaza, & J. Weaver (Eds.), *Posthumanism and educational research* (pp. 17–29). Routledge.

Snaza, N., & Weaver, J. A. (2015). Introduction: Education and the posthumanist turn. In N. Snaza, & J. Weaver (Eds.), *Posthumanism and educational research* (pp. 15–28). Routledge.

Somerville, M. (2017). The Anthropocene's call to educational research. In K. Malone, S. Truong, & T. Gray (Eds.), *Reimagining sustainability in precarious times* (pp. 17–28). Springer.

Steffen, W., Broadgate, W., Deutsch, L., Gaffney, O., & Ludwig, C. (2015). The trajectory of the Anthropocene: The great acceleration. *The Anthropocene Review*, *2*(1), 81–98. https://doi.org/10.1177/2053019614564785.

Stengers, I. (2018). *Another science is possible: A manifesto for slow science* (S. Muecke, Trans). Polity Press.

Stern, D. N. (2010). *Forms of vitality: Exploring dynamic experience in psychology, the arts, psychotherapy, and development*. Oxford University Press.

Taylor, A. (2013). Reconfiguring the natures of childhood. Routledge.

Taylor, A. (2017). Beyond stewardship: Common world pedagogies for the Anthropocene. *Environmental Education Research, 23*(10), 1448–1461. https://doi.org/10.1080/13504622.2017.1325452.

Taylor, A., & Pacini-Ketchabaw, V. (2015). Learning with children, ants, and worms in the Anthropocene: Towards a common world pedagogy of multispecies vulnerability. *Pedagogy, Culture & Society, 23*(4), 507–529. https://doi.org/10.1080/14681366.2015.1039050.

The Invisible Committee. (2014). *To our friends*. MIT Press.

Thunberg, G. (2019a). Greta Thunberg's World Economic Forum 2019 special address. http://opentranscripts.org/transcript/greta-thunberg-world-economic-forum-2019/. Accessed 10 Jan 2020.

Thunberg, G. (2019b). Speech to MPs: You did not act in time. UK: Houses of Parliament. https://www.theguardian.com/environment/2019/apr/23/greta-thunberg-full-speech-to-mps-you-did-not-act-in-time. Accessed 10 Jan 2020.

Todd, Z. (2015). Indigenizing the Anthropocene. In H. Davis, & E. Turpin (Eds.), *Art in the Anthropocene: Encounters among aesthetics, politics, environments and epistemologies* (pp. 241–254). Open Humanities Press.

Trafí-Prats, L. (2020). Fugitive pedagogies: Decolonising black childhoods in the anthropocene. *Discourse: Studies in the Cultural Politics of Education, 41*(3), 359–371. https://doi.org/10.1080/01596306.2019.1589671.

Tsing, A. (2015). *The mushroom at the end of the world: On possibility of lie in capitalist ruins*. Princeton University Press.

Turpin, E., & Davis, H. (2015). Art & death: Lives between the fifth assessment & the sixth extinction. In H. Davis, & E. Turpin (Eds.), *Art in the anthropocene: Encounters among aesthetics, politics, environments and epistemologies* (pp. 3–29). Open Humanities Press.

Weheliye, A. G. (2014). *Habeas viscus: Racializing assemblages, biopolitics, and black feminist theories of the human*. Duke University Press.

Whitehead, A. N. (1968). *Modes of thought*. The Free Press.

Whitehead, A. N. (1978). *Process and reality*. The Free Press.

Williams, R. (1977). *Structures of feeling. Marxism and literature* (pp. 128–135). Oxford University Press.

Winnicott, D. W. (1989). *Playing and reality*. Routledge.

World Wildlife Fund (2014). Living planet report 2014. https://www.worldwildlife.org/pages/living-planet-report-2014. Accessed 5 Jan 2020.

Wynter, S., & McKittrick, K. (2015). Unparalleled catastrophe for our species? Or, to give humanness a different future: Conversations. In McKittrick, K. (Ed.), *Sylvia wynter: On being human as praxis* (pp. 9–89). Duke University Press. https://doi.org/10.1215/9780822375852.

Youdell, D. (2017). Bioscience and the sociology of education: The case for biosocial education. *British Journal of Sociology of Education, 38*(8), 1273–1287. https://doi.org/10.1080/01425692.2016.1272406

Yusoff, K. (2018). *A billion black Anthropocenes or none*. University of Minnesota Press.

2

CRITICAL CLIMATE CHANGE EDUCATION IN A POSTHUMAN MILIEU

Re-Thinking Climate Change Education

> I feel like some of the older people of politics and society don't really care about climate change because they aren't going to be around to see the full effect.
>
> **Thomas, age 13, CC+Me Researcher**

Children and young people are growing up in increasingly uncertain and precarious times, as the social, affective, and material effects of global climate change permeate their everyday lives, environments, and communities (Lee, 2013; Malone, 2018; Rousell and Cutter-Mackenzie-Knowles, 2020; Selby and Kagawa, 2010). Each consecutive report released by the Intergovernmental Panel on Climate Change (IPCC) has confirmed the increasing extent and impact of anthropogenic climate change at the planetary scale (IPCC, 2007, 2014, 2019). Rapid technological advances in environmental sensing, satellite imaging, and computational modelling are enabling scientists to track and predict the effects of human-induced climatic changes with increasing clarity and precision (Gabrys, 2016). This increasing granularity of scientific climate data has also brought issues of climate justice to the fore, as climate change is already impacting most severely on Indigenous people and children in majority world countries[1] with relatively low ecological footprints (Crate and Nuttall, 2009). Climate change and associated geopolitical issues are currently driving the largest forced migration of human populations in recorded history, as numerous areas of the planet become uninhabitable "dead zones" incapable of sustaining human life (Baldwin and Bettini, 2017).

Cultural issues associated with climate change have also reached a point of saturation within the public domain through diverse forms of informational, digital,

DOI: 10.4324/9781003336006-2

and social media (Lowe et al., 2006; Stokols et al., 2009). Many children and young people are confronted on a daily basis with unsustainable patterns of human consumption, population growth, waste production, habitat destruction, pollution, and contamination that exceed the carrying capacities of the Earth's ecological systems (McNeill and Engelke, 2014). Youth-led climate change activism campaigns, such as SchoolStrike4Climate and Extinction Rebellion, are also bringing children and young people's attention to climate change through social media and other online platforms.

It has been widely acknowledged that innovative and effective forms of climate change education are needed for children and young people worldwide, who will be forced to grapple with the uncertainties of climatic instability brought forth by previous generations (Selby and Kagawa, 2010). Climate change education, however, remains a relatively nascent and under-theorised area of inquiry as considered independently from more established fields such as environmental education; science, technology, engineering and mathematics (STEM) education; and education for sustainability (Læssøe et al., 2009). In our review of existing climate change education research (Rousell and Cutter-Mackenzie-Knowles, 2020), a large number of studies tended to focus on children and young people's scientific knowledge of climate change, often describing it as limited, erroneous, and highly influenced by mass media (e.g. Shepardson et al., 2009).

One of the most common misconceptions acknowledged by multiple studies is the posing of a cause–effect relationship between the depletion of the ozone layer and warming associated with the greenhouse effect (Liarakou et al., 2011). A variety of strategies have been proposed to increase the scientific knowledge of both teachers and their students, including the use of digital media to model climate change processes and concepts (Makrakis et al., 2012; Snow and Snow, 2010), the use of narrative to promote environmental empathy and literacy (McKnight, 2010), and participatory action-research which empowers both teachers and students (Feierabend and Eilks, 2011).

Our review of climate change education research also identified a large number of publications within the subcategory of environmental education (EE). These included publications which made direct reference to environmental education, or any of the associated fields of education for sustainability (EFS), education for sustainable development (ESD), or sustainability education. These publications tended to take a different perspective from the STEM education literature, due to the conceptual and methodological diversity that characterises research in environmental education. While there remains a unifying directive within environmental education to encourage the development of pro-environmental behaviours and values (Pruneau et al., 2006), more instrumental agendas have taken hold in the sub-disciplines of ESD and EFS. Within the field of environmental education, a pro-environmental behaviour is generally taken to be one which minimises the negative impacts of humans on social and ecological systems, or a behaviour which takes steps towards more equitable relationships between humans and their environments (Kollmuss and Agyeman, 2002; Pruneau et al., 2006).

One of the major shifts in the environmental education literature associated with climate change education is towards more holistic approaches to effecting behaviour change, with an increased emphasis on situational and affective influences rather than cognitive and knowledge-based influences (Devine-Wright et al., 2004). Research has indicated that many people who are involved in substantial and sustained environmental action tend to have low levels of scientific knowledge about the environment (Kempton et al., 1995), with increased scientific knowledge about climate change showing little to no correlation with pro-environmental attitudes or behaviour (Dijkstra and Goedhart, 2012).

Within the field of environmental education, we find considerable tension around the emergence of climate change education (CCE) as a standalone field alongside EFS and ESD, among other sub-disciplines of environmental education. Selby and Kagawa (2010), for example, advocate for climate change education as a much-needed antidote to the managerial and instrumental agendas that have overtaken the field of sustainability education. Anderson (2012), on the other hand, argues that climate change education should be comprehensively grounded within an ESD approach. Such an approach would involve the direct participation of children and young people as agents of change within their communities, and also establish strong partnerships between education policymakers and climate scientists to inform climate change curricula with scientific expertise (p. 194).

In their cross-national report on climate change education, Læssøe et al., (2009, p. 15) find that climate change has been a central facet of environmental education in recent decades, but has yet to be established as an independent field of practice and inquiry in many countries. The authors report that climate change education has emerged as a standalone field in the last few years, notably through government-sponsored initiatives in China, Canada, and Denmark, and non-government initiatives in Brazil (p. 15). Drawing on national reports on climate change education across ten nations, Læssøe et al., (2009, p. 16) predict three possible scenarios for the global proliferation of climate change education: (1) that climate change education will develop independently of ESD and environmental education more broadly, and become a substantial component of science education programs, as has been the tendency in the United States and China; (2) that climate change education will develop into an integral component of ESD, adopting a truly interdisciplinary approach to education with a wide variety of variations and interpretations, as has been the case in Australia, the UK, South Korea, and Singapore; and (3) that a hybrid of the first two scenarios will emerge, in which climate change education will be treated as an independent field of practice and inquiry which is related, and yet distinct, from ESD and other subsets of environmental education, while also operating outside the confines of general science education.

Given the predominance of research in science education and environmental education, we were surprised to find the literature on climate change in teacher education to be relatively limited. Most teacher education studies focused on the climate change knowledge of pre-service teachers. In a study of 172 pre-service primary school teachers in Greece, Papadimitriou (2004) found that most believed that climate change was happening based on their experiences outside the classroom, but that very few were aware of actions or strategies for mitigating the effects of climate change. He suggests that innovative resources and instructional techniques are needed which reflect the uncertain, tenuous, and interdisciplinary approaches associated with social studies of science and technology (see, for example, Latour, 2004). Çelikler and Kara (2011) also provide a study which targets the climate change misconceptions of pre-service chemistry and biology teachers in Turkey, also highlighting the common lack of knowledge regarding strategies for mitigation. O'Gorman and Davis (2013, p. 780) argue that teacher education faculties can play a significant role in climate change education not only through student learning but also through their societal connections within the broader educational community. However, they find that in the Australian context many teacher education faculties have been reticent to take up these crucial challenges and opportunities.

More recently, an emerging area of climate justice research has taken shape around studies of youth climate activism across the fields of education and youth studies (Mayes & Center, 2022; Stapleton, 2019; Vaghri, 2018). This work has tended to focus on the new youth politics arising from climate strikes that position climate change as an issue of social and intergenerational injustice (Han and Ahn, 2020; O'Brien et al., 2018), and broadly aims to counter pervasive discourses of children and young people as politically marginalised and disengaged (Cocco-Klein and Mauger, 2018; Phillips et al., 2019; Ritchie, 2021). Josefsson and Wall (2020, p. 1044) suggest that new theorisations of global justice are needed that take this new youth politics into account, arguing that "failure to attend to the experiences and struggles of the young, in all their global diversity and marginalisation from power, leaves global justice theorisation limited and flawed". Ritchie (2021, p. 53) makes a similar argument, but with a focus on young Indigenous activist collectives currently pushing for climate justice. She argues that the current uprising of youth climate activism worldwide can be considered a joining of forces with the sustained activism of Indigenous peoples who have actively resisted the expropriation of environmental resources since colonisation. Children and young people's political investment in climate activism can, in this sense, be located in relation to long-standing political movements for both Indigenous sovereignty and children's rights. Climate justice approaches thus intersect closely with calls for new forms of global citizenship education, which not only question "how young people can be included in decision making", but also "how they can dissent from prevailing norms, lifestyles, decisions, and actions that perpetuate business as usual" (O'Brien et al., 2018, p. 42).

Political and Moral Contortions

> Only people that care can help. Never give up.
>
> **Ella, 9-year-old CC+Me Researcher**

The complex entanglement of climate facts, values, and concerns has become particularly salient over the last ten years through the political and moral manipulation of climate change as an issue of public debate, including widespread denial of climate science as the basis for reliable facts. Moralising and politicising discourses around climate change have become part of children and young people's everyday lives in many parts of the world, ranging from a rebellious politics of dissent to a reactionary politics of denial. Children's immersion in a wide range of cultural and religious configuration also intersects in complex ways with their political orientations and actions in response to climate change (Hickey-Moody et al., 2021). Recent neoconservative and populist movements in North America, South America, Europe, and Australia are circulating widespread climate denial on moral, religious, and political grounds, as scientific evidence is ignored or contested under the auspices of a "post-truth" regime (Cutter-Mackenzie-Knowles and Rousell, 2020). Children and young people are also increasingly exposed to apocalyptic visions of the disastrous impacts of climate change through the internet, social media, and film-based media, leading to a pervasive sense of crisis and fear (Colebrook, 2014). These competing contortions and entanglements of climate fact, value, and concern are contributing to a state of existential anxiety among children and young people, with a large-scale Australian study finding that 25% of children were afraid that the world would end within their lifetimes (Tucci et al., 2007).

Conflicting macropolitical messages and commitments to climate change have made it increasingly difficult for children and young people to trust political leadership on the issue. In the following quote taken from Obama's (2015) speech on the Paris Climate Agreement, we note the assured sense of triumph, achievement, and consensus around an "enduring agreement" that will benefit not only the individual nations involved but the entire world.

> In my first Inaugural Address, I committed this country to the tireless task of combating climate change and protecting this planet for future generations. Two weeks ago, in Paris, I said before the world that we needed a strong global agreement to accomplish this goal – an enduring agreement that reduces global carbon pollution and sets the world on a course to a low-carbon future. A few hours ago, we succeeded. We came together around the strong agreement the world needed. We met the moment.

Over a decade ago, Kevin Rudd (a former prime minister of Australia) similarly described climate change as "the great moral challenge of our generation" (2007). The impetus for such political statements has been the irrefutable ratification of anthropogenic climate change and its associated impacts on Earth by distinguished scientists worldwide (IPCC, 2019). Notwithstanding, the political landscape shifted in 2016 with the election of Donald Trump as president of the United States. Trump is an overt climate change denier, and the political climate quickly became saturated with Twitter comments such as:

> **Donald J. Trump** @realDonaldTrump 29 Dec 2017
>
> In the East, it could be the COLDEST New Year's Eve on record. Perhaps we could use a little bit of that good old Global Warming that our Country, but not other countries, was going to pay TRILLIONS OF DOLLARS to protect against. Bundle up!
>
> **137,498 replies 66,668 retweets 208,740 likes**

> **Donald J. Trump** @realDonaldTrump 6 Nov 2012
>
> The concept of global warming was created by and for the Chinese in order make U.S. manufacturing non-competitive.
>
> **13,000 replies 105,014 retweets 67,673 likes**

These radically disparate political climate change messages have the potential to be confusing and derailing when they are projected onto the public sphere, causing climate change to take on an abstract quality as an amorphous, ideologically driven issue rather than a material-discursive reality. With the upsurge of neoconservative populism, white supremacy, and reactionary forms of nationalism in the United States, Australia, Brazil, and Europe, citizens who feel dispossessed by the moral superiority of neoliberal politicians (such as Obama and Rudd) have been willing to vote and act against their own self-interests, and, in the case of climate change, against the survival interests of the entire biosphere. Populist knee-jerk reactions against the identity politics and elitist moral discourses of neoliberal politicians in the United States have led to the seemingly inconceivable reality of Trump's election and, subsequently, his withdrawal of the United States from the Paris Climate agreement in 2017. Only two years after Obama's triumphant proclamation of the Agreement as a moral victory that "the world needed" we found the political climate moving in an entirely opposite direction, as disenfranchised populations grasp desperately for any vestige of agency, belonging, and understanding in a "post-truth" world.

Colebrook (2019) argues that the relativistic capacity for manipulating climatological facts and values is built into the modern concept of humanity as a force that composes a phenomenological world in its image.

> As long as one remains within the world politics will take place within the polity, and within already established rules and limits of engagement. The condition for the possibility of a world of fake news and alternative facts is the modern conception of the world and politics as a public sphere of ongoing legitimation.
>
> *(Colebrook, 2019, p. 41)*

By turning climate change into a moral issue, the mass politicisation of climate change has also served to overshadow and even negate the lived realities of those who are (and will be) most affected by environmental catastrophes. While the implication may be that *anyone* can have a moral position on climate change in a "post-truth" political climate, the reality is that only a very small group of privileged individuals actually gain public media attention for their views on the issue. This near-total dissolution of any semblance of "truth" or "sense" in macropolitical public discourse around climate change demonstrates the failure of scientific facts, moralism, or discursive political rhetoric to effectively address its existential threat. Public engagement with climate change can now be seen to take place almost entirely at the level of pre-conscious affect, where climate opinions and actions are predominantly modulated through massified feelings of existential anxiety, fear, and hatred via social media, rather than by means of rational discourse, communication, facts, or data (Massumi, 2017). Denial of climate change is currently disengaged from scientific or discursive debate, operating instead through an affective politics of white supremacy, sexism, and neoliberal economics that is spread primarily through social media campaigns.

Unfortunately, all of these affective, political, and moral contortions of climate change have been occurring *at the same time* as the number of climate change refugees (human and more than human) and climate change disaster events have drastically increased (Baldwin and Bettini, 2017). This means that the lives of children and young people, one third of the world's population, are being grossly overlooked in the moral politicising and affective modulation of climate change. Children and young people's ideas, beliefs, understandings, and experiences of climate change are too easily devalued in a post-truth environment where anybody can have an opinion about anything, but only the opinions of the wealthy, persuasive, and powerful actually have influence.

> The only possible solution is if humankind realises that we must evolve to become more realistic with our lifestyles, opinions, focuses and choices; we must take action towards saving our tainted planet, before your thoughtless destruction of environmental resources has gone so far that there can be no reversal. This issue must become the top priority within every single nation. We must band together to reverse the effects of climate change which we, ourselves have caused.
>
> **Nikki, age 12, CC+Me Researcher**

Ultimately, the capacity for a political response to climate change comes down to matters of thoughtfulness and *care*. Nikki, in the quote above, describes the thoughtlessness that accompanies wanton environmental devastation. As nine-year-old researcher Ella articulates at the beginning of this section, "only people that care can help". If you don't *care* about the suffering of people, about the extinction of animals and plants, if you don't *care* about life, or about the Earth's climate, if you only *care* about money, profit, success, recognition, then you simply cannot help. You are political deadweight, dragging everything down. The question, then, is how we make the people *who do care* more visible and agentic in the macropolitical and micropolitical spheres of public life.

Countering the historical invisibility of children and young people in the political domain, the recent upsurge of youth climate activism has established a powerful counter-platform for youth voices on climate change in mainstream news agencies and social media networks. Fifteen-year-old Swedish activist Greta Thunberg has received widespread international media attention for her radical critiques of governmental and corporate inaction on climate change. In many of her public addresses, Thunberg employs a rhetoric that effectively fuses climate science with moral pleas, political critiques, and ethical polemics. In an interview in the *Guardian*, Thunberg (2019) provided an apocalyptic vision of the near future collapse of human civilisation, which she claims are supported by scientific facts:

> Around the year 2030, we will be in a position where we probably set off an irreversible chain reaction beyond human control, that will most likely lead to the end of our civilization as we know it. That is unless in that time, permanent and unprecedented changes in all aspects of industrialized society have taken place. Including a reduction of our CO2 emissions by at least 50%. And please note that these calculations are depending on inventions that have not yet been invented at scale. Furthermore these scientific calculations do not include most unforeseen tipping points and feed back loops. Nor do these calculations include already locked in warming hidden by toxic air pollution. Nor the aspect of equity, which is absolutely necessary to make the Paris Agreement work on a global scale. And these calculations are not opinions or wild guesses. These projections are backed up by scientific facts, concluded by all nations through the IPCC. So if we are to stay below the 1.5 degrees of warming limit, which is still possible within the laws of physics, we need to change almost everything. We need to start living within the planetary boundaries. This will be a drastic change for many, but not for most.

Here Thunberg makes an affective plea for rationality and ethics underpinned by established scientific consensus. Her differentiation between "the many" and "the most" speaks to her long-standing commitment not to universalise or neutralise the disparities between the lifestyles of Western minority populations and the majority world that lives below the line of abject poverty. Yet Thunberg has herself been the subject of critique from far-right political commentators, who have positioned her as a kind of mascot for radical left political interests. The tabloid Australian newspaper the *Herald Sun*, for example, has published editorials by far-right commentator Andrew Bolt, describing Thunberg as a "freakishly" messianic puppet for a green cult of paranoia that is using false beliefs in climate science to mobilise a social communist agenda (Bolt, 2019). This example demonstrates the radical plasticity of climate change as an issue that can be politically and morally manipulated to serve any interests whatsoever, while also illustrating the rise of fascist world views and public abuse currently being propagated through mass media outlets bankrolled by Australia's fossil fuel industries.

From Scientism to Feminist STS

Climate justice approaches bring due attention to the role of cultural politics in shaping engagements and understandings of climate change education, a field historically dominated by science-based approaches that have been largely ineffectual in changing attitudes and behaviours towards climate change (e.g. Brownlee et al., 2013; Devine-Wright et al., 2004). There is now a robust literature suggesting that cognitive understandings of climate science have shown little to no correlation with changes in environmental attitudes or behaviour in the wider population (e.g. Dijkstra and Goedhart, 2012; Kempton, 1997). Selby and Kagawa (2010) even observe a trend towards climate change scepticism and denial in mainstream science education programs. These authors specifically target science education as complicit in framing climate change "as an issue calling for a scientific or technical fix rather than as a pathology of an ethically numb, inequitable and denatured human condition" (p. 42).

As Brownlee et al., (2013, p. 14) argue in their critical analysis of foundational processes influencing climate change attitudes and actions, climate change education needs to move beyond cognitive and scientific knowledge-based approaches in order to engage with the affective politics of the issue. They suggest that climate change education should become responsive to the affective investments, beliefs, orientations, and situational contexts of specific communities, rather than focusing on what people don't know or understand about climate change. Affective connections can then be made between diverse experiences and information about climate change, including situated encounters with social and ecological systems, scientific data, time-lapse photography, digital simulations, maps, fictional narratives, documentary films, and numerous other modalities.

The Climate Change and Me project sought to explicitly address the need for creative and affective modalities of climate change education beyond the techno-scientific models that have dominated the science education and sustainability fields. Rather than turning to scientists and science teachers as the assumed experts on climate change, we turned to children and young people themselves as experts on the affective impacts of climate change within their lives and communities. This movement reflected our study's close engagement with feminist science and technology studies (STS), which has historically situated science and technology within an expanded field of critical, political, affective, and material-discursive inquiry (see, for instance, Barad, 2007; de La Bellacasa, 2011; Despret, 2013; Haraway, 1992; Harding, 2009; Stengers, 1997; TallBear, 2017). Haraway's (2016) more recent work provides a catalyst in this regard, advocating for the development of a "new new synthesis" between the environmental arts, humanities, and sciences which could tie "together human and nonhuman ecologies, evolution, development, history, affects, performances, technologies, and more" (p. 63).

In accounting for key precursors to the current posthumanist convergence of scholarship across fields, Bignall and Braidotti (2019) describe the significant role of feminist STS scholars in theorising "the complex and multiscalar consequences of productive encounters between differences when they are brought into affective relationships, always in particular and contingent circumstance of engagement" (p. 5). Our engagement with STS scholarship has helped situate our resistance to the lurking atmospheres of scientism in the institutional framing of climate change education, a presence that continues to position scientific knowledge as a more valued source, object, and aim of education than any other practices or forms of knowledge (Somerville and Green, 2012). Our resistance to scientism is not necessarily a resistance to science, but rather, to its claims of mastery and political neutrality. In attempting to bracket out science as a rarefied, universal, and superior knowledge practice dissociated from cultural, social and political contamination, scientism has the paradoxical effect of "shutting one's ears to what is most interesting about science as such" (Morton, 2016, p. 29), namely its capacity to evoke speculation, wonder, creativity, and social transformation. As feminist STS scholars have vigorously demonstrated, acknowledging the political, aesthetic, and speculative dimensions of science does not refute empirical inquiry but, rather, reveals the ways that empirical facts are materially and discursively generated through "a dynamic set of open-ended practices iteratively refined and reconfigured" (Barad, 2007, p. 167).

Belgian philosopher and chemist Isabelle Stengers has been at the forefront of feminist science studies since the early 1970s. Her collaborative research with Nobel Prize–winning scientists such as Illya Prigogine and Barbara McClintock is widely acknowledged for developing relational, situated, and ecological perspectives on the nature of scientific inquiry and knowledge, revealing how science operates within a cosmopolitical ecology of practices and values that include both human and non-human agencies (Stengers, 1997, 2010, 2018). Through the work of Stengers and numerous other feminist scholars, STS has often operated as a powerful ally for scientists by showing how science actually "works" within an expanded political field. Rather than being reducible to a universal and inviolable scientific method, science is resituated within an ecology of practices that works by following, mutating, and potentialising material-discursive forces and agencies, producing contestable facts, evidence, and results through the value-laden iteration of particular practices, apparatuses, and logics (Rousell, 2021). In other words, science, like any other field of practice, cannot be abstracted from its situated political entanglements with the world's becoming (Stengers, 2010).

In the Climate Change and Me study, STS provided a rigorous underpinning for innovative approaches to climate change education which acknowledge that "empirical science and rational discourse are largely continuous with other ways of feeling, understanding, and engaging with the world … [including] art, myth, religion and narrative" (Shaviro, 2015, p. 13). Rather than approaching science as a system of structural connections between preconstituted terms, we invested in practices and processes that situate science within a more-than-human *political ecology*, or what Stengers terms a "cosmopolitics". Stengers (2010) famously studied the practices of witchcraft using the same cosmopolitical approach she uses to study scientific laboratory practices. While witchcraft and biochemistry involve very different onto-epistemological views of the world, both are shown to be "efficacious" in the fabrication of new entanglements of fact and value according to the specificity of the worldly practices they deploy. Stengers' focus on the specificity of practices constitutes an *ontological pluralism* orientated towards singular, self-causing events whose "natures" are characterised by the relations they form and sustain with other events (Rousell, 2020). Our openness to ontological pluralism realised through the specificity of practices also brings our study into resonance with Indigenous sciences which resist the separation of nature from culture and religion, thus preserving a cosmopolitical continuum between science, art, ritual, and numerous other knowledge practices (Bignall and Rigney, 2019; Cajete, 2006; Rosiek et al., 2020; TallBear, 2017).

Education at the End(s) of the World(s)

> Before the world was ending, the beauty was incomparable.
> But now the perishing heart of the world decays bit by bit.
>
> **Eva, age 10-year-old, CC+Me Researcher**

Our analysis of STS in relation to recent political discourse on climate change in the public domain demonstrates the current blurring between scientific facts and religious beliefs, conservative and radical politics, humanist moral rights, and predatory profiteering. As politics become increasingly driven by affective manipulation through viral memes and other mass forms of social media, a climate of fear and anxiety is displacing the possibility for rational discourse and democratic participation. The living complexity of climate change is not only destabilising the climate, but also the conventional platforms of humanist critique, moral judgement, representational democracy, and identity politics. As Haraway (2016) notes, climate change advances along its own inhuman trajectory of planetary transmutation regardless of humanity's sense of "despair, cynicism, or optimism and the belief/disbelief discourse" (p. 54). As much as this posthuman condition requires new platforms for demanding action on climate change at the macropolitical level, it also involves learning to affectively inhabit a decaying planet while sustaining the micropolitics of a life worth living (Rousell et al., 2022). Put simply, it requires *learning to live at the end(s) of the world(s).*

In many ways, today's children and young people are beginning their lives at the end of a world predicated on a humanist metaphysics of transcendent values and universalising democratic ideals. Without any clear reference point between human and non-human events, between foreground and background, we are now seeing the dissolution of this concept of a phenomenological "world" that has been determined through the (white, male) lens of human events and human history (Colebrook, 2014). While every moment in our daily lives contributes in some way to climate change, the massively distributed and uncontainable event that is "climate change" strikes back with an inhuman force that radically exceeds human perceptions of space, time, and political agency. Climate change does not wait for human polities to manipulate one another while deferring transformative action. As Haraway (2016) warns us, climate change operates through a political ontology that is, quite literally, *other than human* and harbours no concern for human life. Climate change thus signals not only the sixth great extinction event in the Earth's history but also the horizon of extinction of the human species as an exceptional category of worldmakers (Colebrook, 2019). Rather than encountering the Anthropocene as a "human age", it may be that inhuman objects and events are now "responsible for the next moment of human history and thinking" (Morton, 2013, p. 201).

> What relation might we chart between the end of the world in the theoretical sense (where we abandon the conception of an objective world to which language and discourse might be referred), and the end of the world in the ecological sense, where we fear the actual destruction of everything that has composed our reality?
>
> *(Colebrook, 2019, pp. 40–41)*

Colebrook's two senses of the end of the world were reflected throughout the Climate Change and Me project, as children and young people created a range of poetic, visual, and creative accounts of a decaying world increasingly devoid of stable forms of objectivity and life (see Figure 2.1). As Morton (2013, p. 101) argues, thinking and working at the end(s) of the world(s) require "a profound upgrade of

FIGURE 2.1 Photography by Kairo, 12-year-old CC+Me Researcher, with the annotation: "The sun beats down on the dark, dark world. The clouds that gave shelter are gone. There's no life left on the barren black world. Everyone melted so no one can mourn".

our ontological tools" in light of massively distributed geo-traumatic events such as global warming, radioactive contamination, and the mass extinction of plant and animal species. Such unprecedented environmental, scientific, and technological change means that "new sorts of thinking and making are now possible, and called for, in response to new material situations of daily life" (Ellsworth and Kruse, 2012, p. 9). We see this new kind of thinking and making emerging in the work of the CC+Me researchers, as they invoked alternative sensibilities for grappling and coping with the extraordinary changes and threats associated with the climate emergency.

However, it is also important to acknowledge that the induction of a state of crisis through "end of the world" thinking continues to serve the neoliberal agenda of capitalistic extraction, exemption, and control (Colebrook, 2019). Whyte (2018) observes how Western "end of the world" narratives reflect a sense of "white panic" around climate change that is largely out of synch with Indigenous cultures who have lived under the threat of extinction since colonisation. We discuss these concerns in more detail in Chapter 6. As the anonymous collective the *Invisible Committee* emphasises, even the end of the world can serve the dominant interests of the capitalistic-colonial regime.

> It's an endless end, a lasting apocalypse, an indefinite suspension, an effective postponement of the actual collapse, and for that reason a permanent state of exception. The current crisis no longer promises anything; on the contrary, it tends to free whoever governs from every constraint as to the means deployed.
>
> *(The Invisible Committee, 2014, p. 9)*

Indeed, the end of the world can still be extremely profitable for those positioned to exploit it, through schemes ranging from climate engineering to cloning, virtual memory preservation, and elaborate "arks" ensuring the survival of the super-wealthy (Demos, 2018). This lends a sense of urgency to recent calls for new ontological tools and knowledge practices that are responsive to the risk that climate change poses to current social orders and qualities of life. At the very moment that the onset of climate change is apprehended through advanced satellite imaging, geo-chemical testing, and computer simulations, the very concept of life is also being radically reconfigured through a "new milieu of twenty-first-century life sciences, life philosophies, and life politics" (Weinstein and Colebrook, 2017, p. 3). Our project of generating new climate child imaginaries is thus also a project of forming new imaginaries of *life*, as it is precisely at the end(s) of the world(s) signalled by climate change that entirely new understandings and imaginings of life are necessitated.[2]

Learning in Times of Unsustainability

> It is dark on the horizon... but through this project it is becoming clearer. Clearer of the way out of the dark, to the age of natural information.
>
> **Kairo, age 12, CC+Me Researcher**

Perspectives on climate change and extinction from the posthumanities open up new possibilities for research into children and young people's everyday learning experiences in the contemporary moment. Intertwined with the existential threat of climate change, today's children and young people also inhabit an age in which the very nature of life is being reconstituted through postgenomic sciences, ubiquitous computing, and artificial intelligence, such that the boundaries between life and non-life are being recharacterised as plastic, permeable, and open to continuous variation (Braidotti, 2013; Frost, 2016). The cumulative sensing and computational capacities of twenty-first-century media networks have become powerful agencies and elemental components of everyday environments, as mobile phones, satellite navigation systems, digital sensors, the mining of rare-earth minerals, and the ubiquitous data cloud are linked to material, metabolic processes that are enmeshed with the Earth's biosphere and ecological systems (de Freitas et al., 2020; Gabrys, 2016; Hansen, 2015). Bodies, cultures, technologies, and environments are now understood to permeate and interpenetrate one another, with the porous membranes of skin, cell, and touchscreen operating as "fluid mosaics" for the continuous "trafficking" of molecular, biochemical, and semiotic flows (Frost, 2016, p. 27). These radical changes in the nature(s) and understandings of life in the twenty-first century call for a renewed conception of education in response to the rapidly changing "posthuman ecologies" of contemporary life (Braidotti and Bignall, 2019).

As we elaborate throughout the chapters of this book, our work on the CC+Me project connects with the emergence of critical posthumanism in the fields of environmental education and childhood studies over the last two decades, offering new platforms "for an 'unlearning' of anthropomorphic ways of educating about the world" (Malone, 2015, p. 3). In many ways the uptake of posthumanist scholarship has acted like a "storm on thought" in environmental and sustainability education, calling into question historical "presumptions, logics and methods of reason" and forcing both fields to re-evaluate "what *really matters*" (Payne, 2016, p. 169, emphasis in original). A series of contentious and experimental works by Noel Gough (2004, 2006), for example, have explored the ways that posthumanist concepts of the rhizome, becoming, diffraction, and the cyborg can impact on the development of creative pedagogies in environmental education. Weaver (2010, p. 2) has further explored the impact of advances in the biosciences on the field of curriculum studies, drawing on posthuman science fiction narratives that have "prophesised our present condition and forever altered our mind's imagination" in relation to social and environmental issues.

One of the more robust movements in posthumanist education studies has emerged in the area of early childhood education. Drawing on the feminist new materialisms of Barad (2007) and Haraway (2008), Lenz Taguchi (2010, p. 90) has developed the concept and practice of an "intra-active pedagogy" that is predicated on "the force of learning [which] does not separate thinking and bodily doing from objects, matters, time, spaces and places". A group of interdisciplinary scholars have also established the *Common Worlds Childhood Research Collective* to explore children's entangled relations with "other species, the material world and with place" (Taylor et al., 2012; Taylor and Pacini-Ketchabaw, 2015).

These relatively recent engagements with the posthumanities in environmental education and childhood studies can be situated within an extensive genealogy of post-anthropocentric thought across a range of disciplines and fields. Harvester and Blenkinsop (2010, p. 123) highlight the foundational work of ecofeminist thinkers who have contested the patriarchal "logic of domination" over nonhuman animals and environments since the 1970s, while actively promoting "nondualistic and nonhierarchical systems of relations" as alternatives to humanist metaphors, concepts, and understandings of environmental education. Ecofeminist thinkers including Vandana Shiva (Mies and Shiva, 1993), Val Plumwood (1986), and Deborah Bird-Rose (2002) proposed non-anthropocentric conceptualisations of environmental ethics and ecological relations of care that resist deeply entrenched nature/culture binaries inherited from Enlightenment philosophical legacies. As a movement that has explicitly challenged the "masculinist universalism" of technoscientific knowledge production under capitalistic economic regimes, ecofeminist theory is consistently cited as a significant precursor to contemporary studies in critical posthumanism (Braidotti and Bignall, 2019, p. 2). Ecofeminist genealogies have also developed in close historical connection with diverse traditions of Indigenous and non-Western philosophy, which, in a plurality of ways, situate life as "materially embedded in the geoformations and trans-species influences that shape and define existence in relational terms" (p. 2).

There is also a history of work in environmental education that has sustained an engagement with eco-philosophical concepts and approaches to environmental justice since the 1960s, including approaches associated with the deep ecology (Næss and Jickling, 2000) and social ecology movements (Payne, 2010). While environmental education was originally founded as a socio-behavioural extension of environmental science, it also has a minor history of interdisciplinary engagement across the arts, humanities, and sciences that is often overlooked in contemporary posthumanist interventions.

In her historical account of the field, Palmer (2002) describes how environmental education has always been an *inter*disciplinary field, with a genealogy linked to the emergence of environmental studies and the American pragmatist tradition of philosophy in the early twentieth century. Yet the problem of disciplinary hierarchy persists in the field of environmental education, to the extent that the extractive logic of positivism continues to police a dominant regime of white, masculinist thinking and practice. The work of the environmental educator, artist, writer, and theorist continues to be subordinated to scientific methods of categorisation, verification, and logical explication that predetermine *what matters* in environmental education. To this extent, the posthumanist approach to climate change education developed in this book questions whether the humanist legacy and current trajectories of environmental and sustainability education are capable of responding to the "end(s) of the world(s)" articulated in the previous section. Posthumanism offers a crucial intervention into environmental education precisely because it rejects universalist claims of logical positivism operating under Enlightenment ideals of "humanizing" civilization and progress (Cutter-Mackenzie-Knowles and Rousell, 2020; Malone, 2016). By resisting anthropocentric and technocratic "solutions" and embracing more-than-human complexity, climate change education has the capacity to cultivate "an encounter with eco-catastrophe that does not reterritorialise back upon an *all-too-human* image of life' (Wallin, 2015, p. 142, emphasis in original).

Posthuman Ecologies

Plastic everywhere
Obviously all of the animals are sick
Living
Life isn't that great with pollution
Unfortunately people won't stop

Ella, age 9, CC+Me Researcher

In many ways the concept of "sustainability" has created a stumbling block for climate change education initiatives that seek to resist anthropocentric practices rooted in assumptions of human exceptionalism, dominance, and ecological sovereignty (Jickling, 2001, 1994). As Smith (2011, p. xiii) argues, environmental and sustainability education initiatives too often fail to question the anthropocentric assumptions inherent in claims to ecological sovereignty, which "presumes human dominion and assumes that the natural world is already, before any decision is even made, fundamentally a human resource". This entrenched and often intractable anthropocentrism has the stultifying effect of foreclosing "alternative ontological arrangements with the nonhuman and material world" (Lloro-Bidart, 2015, p. 141), inhibiting any significant movement on contemporary social and environmental issues such as "animal rights, the already accomplished ecological catastrophe, bioethics and biotechnology, and the increasing imbrication of human cultures with computers and other machines" (Snaza and Weaver, 2015, p. 7).

Critical posthumanist and post-anthropocentric perspectives are central to this book's re-imagination of climate change education in ways that are politically inclusive of the nonhuman systems on which ecological survival depends, as well as "that which is already inhuman within the human" (Wallin, 2015, p. 142).[3] Unless these concepts are radically re-visioned and re-tooled, it is questionable whether the terms "sustainability" or "sustainable development" can retain any coherent meaning or function at a time that demands a radical reversal of entrenched claims to human superiority and dominance (Braidotti, 2013; Morton, 2013).

Recent developments in feminist new materialisms and critical posthumanism offer a number of ways forward in this regard. Bennett's (2010, p. 111) "vibrant materialism" extends environmental politics to encompass not only humans and other animals but also inorganic "things" such as the power grid and plastic waste, re-positioning sustainability as an affirmation of planetary life that always exceeds the human. Braidotti (2013) similarly proposes a vitalist critique of sustainability that is oriented towards the future co-existence of human and nonhuman life. She describes how the desire for a liveable future can both affirm and catalyse the collective enactment of a liveable present (p. 192). For Braidotti, however, the practical affordances rendered by the concept of sustainability are entirely contingent on the capacity for unfettered creativity, open cultural systems, and visionary modes of thought. "A prophetic or visionary dimension is necessary", Braidotti writes, "in order to secure an affirmative hold over the present, as the launching pad for sustainable becomings or qualitative transformations of negativity and the injustice of the present" (ibid.). In this sense, the posthumanist promise of climate change education lies in its capacity to "palpate a different quality of response" that eschews humanism's "aspirations for transcendence", while affirming the radical potentials for future co-existence (Wallin, 2015, p. 144). From this posthumanist view, it is the assumption that humans are somehow separate from the rest of nature that is inherently flawed and catastrophically unsustainable, and which yet continues to play out through the unquestioned execution of capitalistic and anthropocentric logics of extraction that prey on the field of more-than-human life (Massumi, 2018).

Climate Change Education in an Expanded Field

> Is my occurrence on the face of this planet just to create a life that may support me ... or to create a life for others too?
>
> **Nikki, age 12, CC+Me Researcher**

Throughout this book, we argue for a posthumanist reconceptualisation of climate change education that is in conversation with, and yet distinct from, previous trajectories of research and practice in environmental and sustainability education. We see this new line of inquiry opening up not only in our own projects, but also in a new wave of climate change education research focusing on affective atmospheres and entanglements of climate science, art, media, and technology (Piotrowski, 2019; Verlie, 2019, 2017). In contributing to this new line of work, we suggest that climate change education needs to draw on participatory and creative approaches from multiple disciplines in establishing itself as distinct from science education, sustainability education, and environmental education. The scientific, social, ethical, and political complexities of climate change call for such an approach, which empowers children and young people to meaningfully engage with entanglements of climate fact, value, power, and concern across multiple scales, temporalities, and modes of existence. Drawing on posthumanist perspectives introduced over the course of this chapter, our approach to climate change education in this book is framed by the following key arguments:

- That climate change education begins at the "end(s) of the world(s)" as determined by human events and human history;

- That climate change education demands a new synthesis of critical perspectives across the environmental sciences, technologies, arts, humanities, and social sciences;
- That climate change education is emerging in a post-truth, neoliberal political climate of increasing uncertainty and instability;
- That climate change education is taking shape under the conditions of a technological transmutation of life processes at planetary scale;
- That climate change education challenges anthropocentric concepts of stewardship and sustainability that have historically underpinned environmental and sustainability education programs;
- That climate change education engages critically with the reality of unsustainability and the possibility of extinction, while pursuing a visionary and affirmative agenda for living and learning on a damaged planet; and,
- That climate change education needs to include the perspectives, experiences, and imaginings of children and young people in diversely situated communities and environments, including those of First Nations and Indigenous peoples.

These arguments resituate climate change education in an expanded field of ontological, social, political, technical, and environmental milieus that are shifting rapidly, and increasingly entangled with one another. Based on our experiences working on the *Climate Change and Me* project, we found that these arguments were born out through the inquiry processes that children and young people engaged as thinkers and researchers. Rather than starting from an anthropocentric position of managing natural resources through techno-scientific "solutions" to complex problems, children and young people began with an affective and aesthetic immersion in an age of unsustainability, climate crisis, mass extinction, the erosion of futures, speculative worlds, and more-than-human concerns. This constituted the development of new modes of climate change education which were open to radical and visionary alternatives for future life, necessarily drawing on practices associated with environmental activism, social and political intervention, digital innovation, citizen science, and the creative arts.

In addressing the naming and framing fragmentations of environmental education and its iterations (including education for sustainable development and education for sustainability), the children and young people's research cut through the semantics to speak far more widely than the partiality of "sustainability" or "sustainable development". Overall, we found that none of the researchers mentioned the word sustainability in the workshops, interviews, website, poems, or other creative works generated over the course of the project. While we discussed sustainability at the research workshops and asked the students why they did not use the concept, the CC+Me researchers tended to make comments such as "because it's meaningless", or "sustainable for who?" In lieu of any meaningful concept of sustainability, they gravitated towards dark ecological and vital materialist concepts of co-existence that de-centred human superiority and cultivated affective relationships with the earth's biosphere and climatic systems. To us, this indicates that the concept of sustainability bears little to no traction with children and young people who recognise the Earth's climate change (historically and in the Anthropocene) as an inhuman social and environmental event that threatens human and nonhuman existence. This highlights the problematic nature of education for sustainable development, education for sustainability, and environmental education initiatives which fail to critically engage with the posthuman condition. While these fields historically developed in response to environmentalist, conservationist, and even economic rationalist concerns, climate change education emerges at a time when such humanist ideologies of unmitigated progress and techno-centric management are necessarily being called into question. The question of "education for what?" has becomes a moot point, unless we simply propose that education is *for life*, in all its wild, threatened, and threatening diversity.

If climate change education is to unfold its posthumanist potential we must move beyond the discipline boundaries of environmental education, education for sustainable development and education for sustainability (and other sustainability-education modes). This book argues that climate change education does not simply slip into one of these disciplines or sub-disciplines, but presents an emerging field of practice and research that is directly responsive and accountable to the rapidly changing environmental conditions of Earthly life. As shared throughout the chapters of this book, the findings from the *Climate Change and Me* project suggest that climate change education operates in an expanded field of experience and inquiry that is profoundly different from existing modes of environmental or sustainability education. This entails a radical expansion of the purview of education itself, both in terms of an expanded spatio-temporal scale and complexity of educational thinking and practice, as well as an expanded diversity of actors and forces that participate and contribute to the instantiation of education as a global project. For the young philosophers and researchers we worked with in *Climate Change and Me*, creativity and the arts are crucial to learning to live on a damaged planet and navigate its unique challenges. For ourselves as educators and researchers, we are both invigorated and hopeful that today's children and young people might be key actors in determining the shape of this new field in the making.

Notes

1 The term "majority world" is used to denote non-Western countries (formerly referred to as third world or developing countries). Majority world countries constitute two-thirds of the Earth's human population (Alam, 2008).

2 Weinstein and Colebrook (2017) recontextualise critical studies of climate change and the Anthropocene within the larger interdisciplinary matrix of "critical life studies". This repositioning is helpful for our re-imagining of climate change education because it intersects with numerous other critical projects and studies for whom life has become a framing problem, including affect studies; new vitalisms, materialisms, and empiricisms; decolonial theory; critical disability studies; critical animal studies; transgender and queer

studies; posthumanisms of various kinds; science and technology studies; speculative realisms; and material feminisms (p. 4). These fields are generating new cartographies of life in the twenty-first century by developing ontological, critical, and axiological tools for living and learning on a damaged planet.

3 Here Wallin notes that contemporary biology has revealed that "what we call the human body is in fact composed of 90 percent inhuman organisms" (2015, p. 142).

References

Alam, S. (2008). Majority world: Challenging the West's rhetoric of democracy. *Amerasia Journal, 34*(1), 88–98. https://doi.org/10.17953/amer.34.1.l3176027k4q614v5.

Anderson, A. (2012). Climate change education for mitigation and adaptation. *Journal of Education for Sustainable Development, 6*(2), 191–206.

Baldwin, A., & Bettini, G. (2017). *Life adrift: Climate change, migration, critique*. Rowman & Littlefield.

Barad, K. (2007). *Meeting the universe halfway: Quantum physics and the entanglement of matter and meaning*. Duke University Press.

Bennett, J. (2010). *Vibrant matter: A political ecology of things*. Duke University Press.

Bignall, S., & Braidotti, R. (2019). Posthuman systems. In R. Braidotti, & S. Bignall (Eds.), *Posthuman ecologies: Complexity and process after Deleuze* (pp. 1–16). Rowman and Littlefield.

Bignall, S., & Rigney, D. (2019). Indigeneity, posthumanism and nomad thought: transforming colonial ecologies. In R. Braidotti, & S. Bignall (Eds.), *Posthuman ecologies: Complexity and process after deleuze* (pp. 159–183). Rowman and Littlefield International.

Bird-Rose, D. (2002). *Indigenous ecologies and an ethic of connection. Global ethics and environment* (pp. 189–201). Routledge.

Bolt, A. (2019). The disturbing secret to the cult of Greta Thunberg. https://www.heraldsun.com.au/blogs/andrew-bolt/the-disturbing-secret-to-the-cult-of-greta-thunberg/news-story/55822063e3589e02707fbb5a9a75d4cc. Accessed 16 Jan 2020.

Braidotti, R. (2013). *The posthuman*. Polity Press.

Braidotti, R., & Bignall, S. (2019). *Posthuman ecologies: Complexity and process after deleuze*. Rowman & Littlefield.

Brownlee, M. T. J., Powell, R. B., & Hallo, J. C. (2013). A review of the foundational processes that influence beliefs in climate change: Opportunities for environmental education research. *Environmental Education Research*, 19(1), 1–20. https://doi.org/10.1080/13504622.2012.683389.

Cajete, G. (2006). Western science and the loss of natural creativity. In D. T. Jacobs (Ed.), *Unlearning the language of conquest* (pp. 247–259). University of Texas Press.

Çelikler, D., & Kara, F. (2011). Determining the misconceptions of pre-service chemistry and biology teachers about the greenhouse effect. *Procedia-Social and Behavioral Sciences, 15*, 2463–2470.

Cocco-Klein, S., & Mauger, B. (2018). Children's leadership on climate change: What can we learn from child-led initiatives in the US and the Pacific Islands? *Children, Youth and Environments, 28*(1), 90–103.

Colebrook, C. (2014). *Death of the PostHuman: Essays on extinction* (Vol. 1). Open Humanities Press.

Colebrook, C. (2019). Slavery and the trumpocene: It's not the end of the world. *Oxford Literary Review, 41*(1), 40–50.

Crate, S. A., & Nuttall, M. (2009). *Anthropology and climate change: From encounters to actions*. Routledge.

Cutter-Mackenzie-Knowles, A., & Rousell, D. (2020). The mesh of playing, theorizing, and researching in the reality of climate change: Creating the co-research playspace. In A. Cutter-Mackenzie-Knowles, K. Malone, & E. Barratt Hacking (Eds.), *Research handbook on childhoodnature: Assemblages of childhood and nature research* (pp. 199–222). Springer Nature.

de Freitas, E., Rousell, D., & Jager, N. (2020). Relational architectures and wearable space: Smart schools and the politics of ubiquitous sensation. *Research in Education, 107*(1), 10–32.

de La Bellacasa, M. P. (2011). Matters of care in technoscience: Assembling neglected things. *Social studies of science, 41*(1), 85–106.

Demos, T. J. (2018). To save a world: Geoengineering, conflictual futurisms, and the unthinkable. *E-Flux Journal, 94*, 1–15.

Despret, V. (2013). From secret agents to interagency. *History and Theory, 52*(4), 29–44.

Devine-Wright, P., Devine-Wright, H., & Fleming, P. (2004). Situational influences upon children's beliefs about global warming and energy. *Environmental Education Research, 10*(4), 493–506. https://doi.org/10.1080/1350462042000291029.

Dijkstra, E. M., & Goedhart, M. J. (2012). Development and validation of the ACSI: Measuring students' science attitudes, pro-environmental behaviour, climate change attitudes and knowledge. *Environmental Education Research, 18*(6), 733–749. https://doi.org/10.1080/13504622.2012.662213.

Ellsworth, E., & Kruse, J. (2012). Introduction. In E. Ellsworth, & J. Kruse (Eds.), *Making the geologic now: Responses to material conditions of contemporary life* (pp. 5–26). Punctum Books.

Feierabend, T., & Eilks, I. (2011). Teaching the societal dimension of chemistry using a socio-critical and problem-oriented lesson plan based on bioethanol usage. *Journal of Chemical Education, 88*(9), 1250–1256.

Frost, S. (2016). *Biocultural creatures: Toward a new theory of the human*. Duke University Press.

Gabrys, J. (2016). *Program earth: Environmental sensing technology and the making of a computational planet*. University of Minnesota Press.

Gough, N. (2004). RhizomANTically becoming-cyborg: Performing posthuman pedagogies. *Educational Philosophy and Theory, 36*(3), 253–265.

Gough, N. (2006). Shaking the tree, making a rhizome: Towards a nomadic geophilosophy of science education. *Educational Philosophy and Theory, 38*(5), 625–645.

Han, H., & Ahn, S. W. (2020). Youth mobilization to stop global climate change: Narratives and impact. *Sustainability, 12*(10), 4127.

Hansen, M. B. (2015). *Feed-forward: On the future of twenty-first-century media*. University of Chicago Press.

Haraway, D. (1992). The promises of monsters: A regenerative politics for inappropriate/d others. In L. Grossberg, C. Nelson, & P. Treichler (Eds.), *Cultural studies* (pp. 295–337). Routledge.

Haraway, D. (2008). *When species meet*. University of Minnesota Press.

Haraway, D. (2016). *Staying with the trouble: Making kin in the Chthulucene*. Duke University Press.

Harding, S. (2009). Postcolonial and feminist philosophies of science and technology: Convergences and dissonances. *Postcolonial Studies, 12*(4), 401–421.

Harvester, L., & Blenkinsop, S. (2010). Environmental education and ecofeminist pedagogy: Bridging the environmental and the social. *Canadian Journal of Environmental Education (CJEE), 15*, 120–134.

Hickey-Moody, A., Cutter-Mackenzie-Knowles, A., Rousell, D., & Hartley, S. (2021). Children's Carbon Cultures. *Cultural Studies ↔ Critical Methodologies, 21*(3), 214–224.

Intergovernmental Panel on Climate Change (2007) '"Climate change 2007. Synthesis Report." Contribution of Working Groups I, II and III to the Fifth Assessment Report of the Intergovernmental Panel on Climate Change'. Geneva, Switzerland: IPCC.

Intergovernmental Panel on Climate Change (2014) '"Climate Change 2014: Synthesis Report." Contribution of Working Groups I, II and III to the Fifth Assessment Report of the Intergovernmental Panel on Climate Change'. Geneva, Switzerland: IPCC.

Intergovernmental Panel on Climate Change (2019) *IPCC Special Report on the Ocean and Cryosphere in a Changing Climate*. Geneva, Switzerland: IPCC.

Jickling, B. (1994). Why I don't want my children to be educated for sustainable development. *Trumpeter, 11*(3), 114–116.

Jickling, B. (2001). Environmental thought, the language of sustainability, and digital watches. *Environmental Education Research*, 7(2), 167–180.

Josefsson, J., & Wall, J. (2020). Empowered inclusion: Theorizing global justice for children and youth. *Globalizations, 17*(6), 1043–1060.

Kempton, W. (1997). How the public views climate change. *Environment: Science and Policy for Sustainable Development, 39*(9), 12–21. https://doi.org/10.1080/00139159709604765.

Kempton, W., Boster, J. S., & Hartley, J. A. (1995). *Environmental values in American culture*. The MIT Press.

Kollmuss, A., & Agyeman, J. (2002). Mind the gap: why do people act environmentally and what are the barriers to pro-environmental behavior? *Environmental education research, 8*(3), 239–260.

Læssøe, J., Schnack, K., Breiting, S., Rolls, S., Feinstein, N., & Goh, K. C. (2009). *Climate change and sustainable development: The response from education*. International Alliance of Leading Educational Institutes.

Latour, B. (2004). *Politics of nature*. Harvard University Press.

Lee, N. (2013). *Childhood and biopolitics: Climate change, life processes and human futures*. Palgrave Macmillan.

Lenz Taguchi, H. (2010). *Going beyond the theory/practice divide in early childhood education: Introducing an intra-active pedagogy*. Routledge.

Liarakou, G., Athanasiadis, I., & Gavrilakis, C. (2011). What Greek secondary school students believe about climate change? *International Journal of Environmental and Science Education, 6*(1), 79–98.

Lloro-Bidart, T. (2015). A political ecology of education in/for the Anthropocene. *Environment and Society*, 6(1), 128–148.

Lowe, T., Brown, K., Dessai, S., de França Doria, M., Haynes, K., & Vincent, K. (2006). Does tomorrow ever come? Disaster narrative and public perceptions of climate change. *Public Understanding of Science, 15*(4), 435–457.

Makrakis, V., Larios, N., & Kaliantzi, G. (2012). ICT-enabled climate change education for sustainable development across the school curriculum. *Journal of Teacher Education for Sustainability, 14*(2), 54–72.

Malone, K. (2015). Posthumanist approaches to theorising children's human-nature relations. In K. Nairn, P. Kraftl, & T. Skelton (Eds.), *Space, Place and Environment, Geographies of Children and Young People* (pp. 185–206). Springer.

Malone, K. (2016). Reconsidering children's encounters with nature and place using posthumanism. *Australian Journal of Environmental Education, 32*(1), 42–56.

Malone, K. (2018). *Children in the Anthropocene: Rethinking sustainability and child friendliness in cities*. Palgrave Macmillan Publishers.

Massumi, B. (2017). *The principle of unrest*. Open Humanities Press.

Massumi, B. (2018). *99 Theses on the revaluation of value: A postcapitalist manifesto.* Minnesota University of Minnesota Press.

Mayes, E., & Center, E. (2022). Learning with student climate strikers' humour: towards critical affective climate justice literacies. *Environmental Education Research*, 1–19.

McKnight, D. M. (2010). Overcoming "ecophobia": fostering environmental empathy through narrative in children's science literature. *Frontiers in Ecology and the Environment, 8*(6), e10–e15.

McNeill, J. R., & Engelke, P. (2014). *The great acceleration: An environmental history of the Anthropocene since 1945.* Harvard University Press.

Mies, M., & Shiva, V. (1993). *Ecofeminism.* Zed Books.

Morton, T. (2013). *Hyperobjects: Philosophy and ecology after the end of the world.* University of Minnesota Press.

Morton, T. (2016). *Dark ecology: For a logic of future coexistence.* Columbia University Press.

Næss, A., & Jickling, B. (2000). Deep ecology and education: A conversation with Arne Naess. *Canadian Journal of Environmental Education (CJEE), 5*(1), 48–62.

O'Gorman, L., & Davis, J. (2013). Ecological footprinting: Its potential as a tool for change in preservice teacher education. *Environmental Education Research, 19*(6), 779–791.

Obama, B. (2015). Statement by the President on the Paris Climate Agreement. https://obamawhitehouse.archives.gov/the-press-office/2015/12/12/statement-president-paris-climate-agreement. Accessed 10 Jan 2020.

O'Brien, K., Selboe, E., & Hayward, B. M. (2018). Exploring youth activism on climate change. *Ecology and Society, 23*(3).

Palmer, J. (2002). *Environmental education in the 21st century: Theory, practice, progress and promise.* Routledge.

Papadimitriou, V. (2004). Prospective primary teachers' understanding of climate change, greenhouse effect, and ozone layer depletion. *Journal of Science Education and Technology, 13*(2), 299–307.

Payne, P. (2010). Moral spaces, the struggle for an intergenerational environmental ethics and the social ecology of families: An 'other' form of environmental education. *Environmental Education Research, 16*(2), 209–231.

Payne, P. (2016). What next? Post-critical materialisms in environmental education. *The Journal of Environmental Education, 47*(2), 169–178. https://doi.org/10.1080/00958964.2015.1127201.

Phillips, L. G., Ritchie, J., & Perales, F. (2019). Surveying support for child and youth political participation in Australia and New Zealand. *Citizenship Studies, 23*(5), 460–485.

Piotrowski, M. (2019). 'An atmosphere, an air, a life': Deleuze, elemental media, and more-than-human environmental subjectification and education. *Environmental Education Research*, 1–12.

Plumwood, V. (1986). Ecofeminism: An overview and discussion of positions and arguments. *Australasian Journal of Philosophy, 64*(sup1), 120–138.

Pruneau, D., Doyon, A., Langis, J., Vasseur, L., Ouellet, E., McLaughlin, E., et al. (2006). When teachers adopt environmental behaviors in the aim of protecting the climate. *The Journal of Environmental Education, 37*(3), 3–12.

Ritchie, J. (2021). Movement from the margins to global recognition: climate change activism by young people and in particular indigenous youth. *International Studies in Sociology of Education, 30*(1–2), 53–72.

Rosiek, J. L., Snyder, J., & Pratt, S. L. (2020). The new materialisms and Indigenous theories of non-human agency: Making the case for respectful anti-colonial engagement. *Qualitative Inquiry*, 26(3–4), 331–346.

Rousell, D. (2020). Cosmopolitical encounters in environmental education: Becoming-ecological in the intertidal zones of Bundjalung National Park. *The Journal of Environmental Education, 52*(2), 133–148.

Rousell, D. (2021). *Immersive cartography and post-qualitative inquiry: A speculative adventure into research-creation*. Routledge.

Rousell, D., & Cutter-Mackenzie-Knowles, A. (2020). A systematic review of climate change education: giving children and young people a 'voice' and a 'hand' in redressing climate change. *Children's Geographies, 18*(2), 191–208. https://doi.org/10.1080/14733285.2019.1614532.

Rousell, D., Ryan, E., Bauer-Nilsen, B., Lai, R. (2022). Animist pedagogies and the endings of worlds: Rituals for the pluriverse. In Burnard, P., MacKinlay, L., Rousell, D., & Dragovic, T. (Eds.) *Doing rebellious research in and beyond the academy* (pp. 263–281). Brill.

Rudd, K. (2007). Climate Change. http://kevinrudd.com/about-2/climate-change/. Accessed 20 Nov 2019.

Selby, D., & Kagawa, F. (2010). Runaway climate change as challenge to the 'closing circle' of education for sustainable development. *Journal of Education for Sustainable Development, 4*(1), 37–50. https://doi.org/10.1177/097340820900400111.

Shaviro, S. (2015). *Discognition*. Repeater Books.

Shepardson, D. P., Niyogi, D., Choi, S., & Charusombat, U. (2009). Seventh grade students' conceptions of global warming and climate change. *Environmental Education Research, 15*(5), 549–570. https://doi.org/10.1080/13504620903114592.

Smith, M. (2011). *Against ecological sovereignty: Ethics, biopolitics, and saving the natural world*. University of Minnesota Press.

Snaza, N., & Weaver, J. A. (2015). Introduction: Education and the posthumanist turn. In N. Snaza, & J. Weaver (Eds.), *Posthumanism and educational research* (pp. 15–28). Routledge.

Snow, B., & Snow, D. (2010). *Birds and berries*. A&C Black Publishers.

Somerville, M., & Green, M. (2012). Place and Sustainability Literacy in Schools and Teacher Education. Paper presented at the Paper presented at AARE, Sydney, NSW, 2–6 December 2012.

Stapleton, S. R. (2019). A case for climate justice education: American youth connecting to intragenerational climate injustice in Bangladesh. *Environmental Education Research, 25*(5), 732–750.

Stengers, I. (1997). *Power and invention: Situating science* (trans: P. Bains). University of Minnesota Press.

Stengers, I. (2010). *Cosmopolitics* (trans: R. Bononno). University of Minnesota Press.

Stengers, I. (2018). *Another science is possible: A manifesto for slow science* (trans: S. Muecke). Polity Press.

Stokols, D., Misra, S., Runnerstrom, M. G., & Hipp, J. A. (2009). Psychology in an age of ecological crisis: From personal angst to collective action. *American Psychologist, 64*(3), 181.

TallBear, K. (2017). Beyond the life/not life binary: A feminist-indigenous reading of cryopreservation, interspecies thinking and the new materialisms. In J. Radin & E. Kowal, (Eds.), *Cryopolitics: Frozen life in a melting world* (179–202). The MIT Press.

Taylor, A., & Pacini-Ketchabaw, V. (2015). Learning with children, ants, and worms in the Anthropocene: towards a common world pedagogy of multispecies vulnerability. *Pedagogy, Culture & Society*, 23(4), 507–529. https://doi.org/10.1080/14681366.2015.1039050.

Taylor, A., Pacinini-Ketchabaw, V., & Blaise, M. (2012). Children's relations to the more-than-human world. *Contemporary Issues in Early Childhood, 13*(2), 81–85.

The Invisible Committee. (2014). *To our friends*. MIT Press.

Thunberg, G. (2019). Greta Thunberg, schoolgirl climate change warrior: 'Some people can let things go. I can't'. https://www.theguardian.com/world/2019/mar/11/greta-thunberg-schoolgirl-climate-change-warrior-some-people-can-let-things-go-i-cant. Accessed 15 Jan 2020.

Trump, D., J [@realDonaldTrump] (2012, November 6). The concept of global warming was created by and for the Chinese in order make U.S. manufacturing non-competitive. [Tweet]. https://twitter.com/realDonaldTrump/status/265895292191248385. Accessed 10 Jan 2020.

Trump, D., J [@realDonaldTrump]. (2017, December 29). In the East, it could be the COLDEST New Year's Eve on record. Perhaps we could use a little bit of that good old Global Warming that our Country, but not other countries, was going to pay TRILLIONS OF DOLLARS to protect against. Bundle up! [Tweet]. https://twitter.com/realDonaldTrump/status/946531657229701120. Accessed 10 Jan 2020.

Tucci, J., Mitchell, J., & Goddard, C. (2007). *Children's fears, hopes and heroes*. Australian Childhood Foundation.

Vaghri, Z. (2018). Climate change, an unwelcome legacy: The need to support children's rights to participate in global conversations. *Children, Youth and Environments*, 28(1), 104–114.

Verlie, B. (2017). Rethinking climate education: Climate as entanglement. *Educational Studies*, *53*(6), 560–572.

Verlie, B. (2019). "Climatic-affective atmospheres": A conceptual tool for affective scholarship in a changing climate. *Emotion, Space and Society*, *33*, 100623.

Wallin, J. J. (2015). Dark posthumanism, unthinking education, and ecology at the end of the Anthropocene. In N. Snaza, & J. Weaver (Eds.), *Posthumanism and educational research* (pp. 172–183). Routledge.

Weaver, J. (2010). *Educating the Posthuman*. Sense Publishing.

Weinstein, J., & Colebrook, C. (2017). Introduction: Critical life studies and the problems of Inhuman rites and posthumous life. In J. Weinstein, & C. Colebrook (Eds.), *Posthumous Life: Theorising beyond the posthuman* (pp. ix–xxvii). Columbia University Press.

Whyte, K. P. (2018). Indigenous science (fiction) for the Anthropocene: Ancestral dystopias and fantasies of climate change crises. *Environment and Planning E: Nature and Space*, *1*(1–2), 224–242.

3

RESEARCH PLAYSPACES

Climate Child Researchers

> When I was told that they wanted kids to research climate change, I was in disbelief. I never really thought that, as a 12 year-old girl living in a small town in the Northern Rivers, I would have that much effect on the world we live in. So when my friends and I began to research climate change, interview our communities and create artworks and videos together as a result, it was quite eye-opening to realize what we, or to be more specific, *I* could do to help our dying world.
>
> **Alita, age 12, CC+Me Researcher**

Child Researchers

In our systematic review of climate change education research, we found that a very minor contingent of the literature has been orientated towards child-framed approaches to climate change education (Rousell and Cutter-Mackenzie-Knowles, 2020a). Only a small selection of papers could be found which drew upon the unique perspectives and experiences of children and young people to inform new frameworks and methods for teaching and learning about climate change (see, for instance, Lawler and Patel, 2012; Tanner, 2010). The Climate Change and Me project aimed to establish a collaborative platform for children to explore new trajectories of inquiry into climate change and the Anthropocene, building on their own experiential knowledges, divergent lines of interest, and emergent creative practices. This was a risky experiment fraught with uncertainty. There was no way to know in advance how children would take up, challenge, and reroute the project in their own ways, or how the dynamic collaborations between ourselves and the children in the project would be negotiated. In many ways we found that our project was flying blind, and that new theoretical concepts and creative methods of

DOI: 10.4324/9781003336006-3

inquiry had to be invented and improvised on the spot as our collaborations with the child researchers unfolded.

The idea of child researchers is not a new concept, with critical debates regarding the conceptualisation, status, and social construction of children's roles in research stretching back several decades (Punch, 2002). There is now an established body of work which works with concepts of "children as researchers" or "children as co-researchers" (Kellett 2005), a field which has its roots in participatory research methodologies and the new sociology of childhood movement promoting the active involvement of children and young people in research (Barratt Hacking et al., 2013). These concepts also coincide with the Convention on the Rights of the Child (UNICEF, 1989), which provoked a renewed consideration of children's rights, with 196 countries currently committed to the convention with the noted exception of the United States.

However, it was not until the early 2000s that the extensive development of explicitly "child-framed" research methodologies began to emerge (c.f. Bell, 2008; Christensen and James, 2000; Kellett, 2005, 2010; Morrow, 2008; Skelton, 2008). A resounding tenet of child-framed research methodologies has been the positioning of the child as an agential subject, rather than a passive object of research (c.f. Sorin and Galloway, 2006). Kellett (2005, p. 2) has argued that children and young people should be "acknowledged as experts on their own lives and if adults genuinely want to understand children and childhood, better ways to seek out child perspective and unlock child voice must be sought". Such calls have intersected with the development of pupil/student voice, photovoice and videovoice as methods that foreground children's subjective positioning and agencies within the context of social research (Cook-Sather, 2006, 2007; Fielding, 2004). However, the vast majority of research described as "child-framed" continues to be adult-initiated and determined predominantly by adult agendas and intellectual framings (Wake, 2008). With the constraints of ethics approval processes and impoverished funding conditions, undertaking research that is authentically child-initiated is inherently risky and incredibly complicated.

While child-framed research methodologies have become more visible in educational and social research more broadly, such methodologies have largely sought to "humanise" the child according to normative hierarchies of knowledge and sociocultural images of the human "subject". As we argued in Chapter 1 with reference to the work of Rolnik (2017) and Guattari (2008), macropolitical attempts to include and empower marginalised voices often serve to reinscribe those same voices within existing regimes of subjectification, essentially attempting to "raise up" those marginalised voices to the perverse image of the "human" or the "social actor" that previously excluded them. As Rolnik (2017) argues, it is precisely the category of the "human subject" that needs to be radically challenged, expanded, and reconnected at the level of micropolitical intensities and more-than-human life. She describes how leftist attempts to assimilate the Indigenous peoples in Brazil through strategies of "mere inclusion" has failed to acknowledge the more-than-human subjectivity "outside the subject" which forms the basis for Indigenous lifeways.

> This focus on mere inclusion suggests to us that left-wing not only assumes the dominant mode of existence as its reference, but also considers it as "the" sole and universal reference, denying any alterity. The consequence is that they lose the crucial opportunity to inhabit the relational fabric woven by these different modes of existence and, above all, to sustain its possible shifting effects that could render void the dominant cartography.
>
> *(Rolnik, 2017, p. 3)*

In this respect, raising children up to the status of rational, responsibilised "human subjects" and "social actors" only serves to reinforce artificial separations of children's lives and political agency from the more-than-human world which is its condition for existence.

It has only been in the last decade that researchers working across childhood and nature studies (now referred to as childhoodnature, as coined in Cutter-Mackenzie-Knowles et al., 2020) have begun to theorise their work with child researchers in terms of more-than-human assemblages which destabilise the figure of the human "subject" as the primary arbiter of experience, knowledge, social agency, inquiry, and value (c.f. Rousell and Cutter-Mackenzie-Knowles, 2020b). This work has challenged conventional framings of "children as researchers" or "children as co-researchers" where the preposition "*as*" invokes an equivalency or "raising up" of the child to the status of the rational, liberal, academic subject. This discourse of equivalency invokes a lesser, postured, and othered identitarian positioning, diminishing the liveliness and anarchic resistance of young people's thought and experience outside the discursive regime of the liberal subject. It also tends to position young people solely as academic knowledge producers, rather than as thinkers and researchers who are *always already* participating in more-than-human ecological systems and aesthetic life processes in dynamic and irreducible ways (Rousell and Cutter-Mackenzie-Knowles, 2020b). In our study, we proposed the term "child researcher" as a simple attempt to remove the prepositional equivalency encoded in "child as subject", "child as social actor", and "child as researcher" discourses.

Our concept of child researcher is inclusive of the concept of child philosopher or child theorist, an acknowledgement of children's powers of thought that has been much slower to develop in depth and rigour in the literature surrounding child-framed research. The question of whether, and how, children can *do* philosophy has been explored by philosopher and childhood studies scholar Karin Murris for some time. While Murris (2000) notes that this question has been addressed (mostly in the negative) over the centuries, she finds that the potential for children to actively think and experiment with philosophy has only barely been explored, with the assumption that adults are intellectually superior to children being one limiting factor. Through what Murris (2016b) describes as "institutionalised ageism", children continue to be largely inscribed as "educables", incapable of demonstrating political will, anarchism, ethical decision-making, creativity, empirical inquiry, and rational thought. This also serves to re-instate humanist claims to logocentrism, which assumes that only intellectual, cognitive, or rational functions of knowing/

being actually matter, or constitute "proper" philosophical, scientific, or artistic contributions.

As we discuss later in this chapter, intensive research in collaboration *with* children and young people can be made possible through the development of a "research playspace" that supports the cultivation of children's own research interests and creative methods of inquiry. The research playspace operates through micropolitical and emergent relations of care for the ongoing process of inquiry itself, a process that is continuously shared and negotiated between child and adult researchers working through emergent and improvised modalities of collaboration. The playspace is not yoked to conventions of academic knowledge production or outputs but, rather, takes shape through collective practices of artistic and philosophical experimentation with germinal trajectories of inquiry as they emergently unfold. Drawing more so from the practice domains of activism, improvised play, and socially engaged arts rather than the normative conventions of social science research, it is important to consider the radical expansion of what qualitative and post-qualitative inquiry can be(come) when child and adult researchers work creatively together in the research playspace. In order to set the stage for the development of the research playspace as a methodological concept, we briefly track the development of children's play as a research focus before exploring Winnicott's theory of play as an oscillation between intensive and extensive milieus.

Transitional Space

Play theory and research have historically been situated in early childhood education and developmental psychology. The concept of adult play is a developing field of inquiry (Lammes, 2019); however, it has only been in recent times that play as a concept or theory has crossed or transcended disciplinary boundaries. Notwithstanding, Fredrich Froebel (1782–1852) is often portrayed as a seminal theorist in play and is widely known for coining the concept of the "kindergarten" (translation – children's garden). Froebel's ideas still firmly endure in early childhood education, and, like other theorists of experiential education (Dewey, 1938, 1956; Vygotsky, 1986, 1997, 2004), Froebel placed deep importance on self-elected activities where adults seldom, yet delicately, intervene in children's play (c.f. MacRae and MacLure, 2021; Wood and Attfield, 2005). Froebel writes that

> Play is the highest form of level of child development. It is the spontaneous expression of thought and feeling – an express which his [*sic*] life requires. This is the meaning of the word play. It is the purest creation of the child's mind as it is also a pattern and copy of the natural life hidden in man [*sic*] and in all things.
>
> *(Froebel, 1967, p. 83)*

Wood and Attfield (2005) position Froebel, Rousseau, and Dewey as seminal theorists in shifting views of early childhood education with play seen as critical to children's learning and development. However, these theorists often position the child at the centre of learning with the environment as a backdrop, prop, setting, or even a "third teacher", as in the Reggio Emilia tradition (Dodd-Nufrio, 2011). In instrumentalising the environment as the context for children's social, physical, and mental development, play theory has not often considered the child and environment as interpenetrating and mutually entangled events of experience. What seems to be denied by conventional play theories is the radical nature of *the outside*, as a chaosmic milieu of inhuman forces that resist and confound human mastery, but are inherently playful and anarchic in their own way (Guattari, 1995). Conventional play theories seem to suggest that play is a way of mastering the environment, of bending it to the whims and needs of the child's development as prefigured according to the logic of the liberal humanist subject. This raises questions around the non-innocence of play with respect to the linked regimes of global capitalism and colonisation, which are founded on the calculated pursuit of mastery as a universal signifier for "human" progress (Singh, 2017).

To develop a more processual and relational understanding of play that mutually implicates children *in* the environments they co-construct and co-inhabit with many other bodies, we turn initially to Winnicott's concepts of transitional space and transitional objects as developed in his book *Playing and Reality* (1989). Winnicott based his theories on observations of infants as they separated from their mothers and began to experiment with the affordances and constraints of their physical environments. Transitional space, as Winnicott describes it, is a spatio-temporal process that puts intensive and extensive milieus of experience into oscillation through playful inquiry and creative experimentation (Ellsworth, 2005, p. 60). The element of play is central to the concept of transitional space, as it is through play that the passages and movements between public and private worlds of experience can be extended, intensified, and sustained along new trajectories. As Elizabeth Ellsworth explains:

> Winnicott saw [transitional space] as a relation of an unknowable (to itself as well as to others) mind/brain and body "interior" to an unknowable and radically other "exterior", and this transit across the space of difference between inside and outside is transitional because encounters with the "not me" that one finds there and the actions that we take in response to such encounters change both the inside of the self and the outside of the social environment. He saw the convergence of inner and outer events and qualities that inaugurates transitional space as a *transitional phenomenon* because it happens in time, not only in space.
>
> *(Ellsworth, 2005, p. 60)*

Winnicott (1989) also thought of transitional space as *potential space*, akin to what Deleuze and Guattari (1994) describe as the virtual plane of immanence on which new conceptual territories are formed, navigated, and sustained. The virtual is, in this sense, a dimension of emergence, creation, and potential that is "like a field of energies that have not yet been expended, or a reservoir of potentialities that have not yet been tapped" (Shaviro, 2009, p. 35). Massumi (2002) similarly describes transitional space as a relational *field of emergence* through which an event takes shape in the actualisation of its virtual potentialities. This brings the notion of transitional space into close proximity with the Spinozan concept of affect as the capacity or potential to affect and be affected, such that "the body coincides with its transitions and its transitioning with its potential" (p. 15).

In more recent work, Massumi (2018) connects the anarchic power of play more closely with modes of resistance to the capitalist monetisation of experience currently infiltrating all corners of individual and collective life. He describes the need to cultivate

> *creative ways to play the affect/intensity differential* that counter the modes of capture of creative tension that are a part of the contemporary ecology of powers, avoiding in particular the personalising capture of affect in emotion endemic to the capitalist economisation process that goes along with it.
>
> *(2018, p. 92, emphasis in original)*

Massumi offers a strong argument against the "personalisation" of play, which reduces the field of affective potentials and intensities of play down to the impoverished image of the liberal humanist subject, an image all too easily co-opted into the predatory capitalist-colonial regime.

Massumi (2002, 2018), Deleuze and Guattari (1994), and Ellsworth (2005) offer a range of linked understandings and approaches to the concept of play, in many ways allying around the proposition that some manner of collective activation is needed for a field of play to be actualised. In this regard, transitional spaces and phenomena are always potential because "nothing makes them inherently or inevitably transitional" (Ellsworth, 2005, p. 60). While an artist's or a teacher's design may hold within it the possibilities for transitional spaces to emerge, it is only through collective, immersive, and playful engagement that these spaces actually come into being. The pedagogical implication of this, as Ellsworth (2005, p. 32) notes, is that transitional spaces can be designed *for*, but not predetermined or forced into existence:

> [A learning environment] holds the potential to become transitional space when it provides opportunities for us to both act in the world and be acted upon by it- while at the same time offering us the flexible stability we need to risk allowing ourselves to be changed by this interaction.

The qualities and elements that are conducive to unlocking the potential of transitional space are described by Winnicott (1989) in terms of a *good-enough holding environment*, which combines the immersive sensation of being held as a child with the impetus to venture beyond this security and into the unknown. This is akin to Stern's (2010) concept of vitality affects, which he uses to describe specific encounters of relation, such as a touch between mother and child, that activate the affective contours of an experience of aliveness.

This notion of the good enough holding environment is very similar to how we imagine the concept of the *research playspace*. It is a space that effectively holds and fosters the possibility conditions for vitality affects to occur, often by surprise, through improvisational and responsive play. It is a space that collectively affirms life as a creative force. In this sense, there can be no template or blueprint for what a "good enough holding environment" or a "research playspace" should look like. Instead, our task as researchers involves "laying out desired qualities for the design of an environment that will not be complete or realised until and unless its users enter it and find their own uses for it" (Ellsworth, 2005, p. 61). The research playspace is thus conceived as a flexible architecture of engagement composed of various materials, ideas, media, technologies, tools, designs, and principles which children can then assemble into new and unforeseen configurations.

Transitional Objects

The flexible elements that populate a "good enough holding environment" for transitional space are described by Winnicott (1989) as *transitional objects*. Winnicott initially developed the concept to describe the ways that infants use comfort objects such as stuffed animals, blankets, or pacifiers as mobile surrogates for the vitality affects linked to the mother's presence. These objects enable the mobile transition of affective attachments into new states of becoming through environmental experimentation. The transitional object is also closely connected with Winnicott's concept of play, as he writes that the "immensely exciting … thing about playing is always the precariousness of the interplay of … psychic reality and the experience of … actual objects" (Winnicott, 1988, p. 55). The transitional object offers the affective and physical continuity, stability, and security needed to enter into an experimental, open-ended engagement with the unknown as the radical outside of the bounded subject. As Ellsworth (2005, p. 60) further explains, "we use transitional objects to imaginatively put ourselves in a transformative relation with the outside". Winnicott's (1989) work on transitional objects is particularly helpful in considering the oscillation between inner and outer spaces in navigating and sustaining ecological dependency relations.

> With his theory of the transitional object, Winnicott jolted all ponders on human nature into a realization of the never-ending oscillation between the *inner* and *outer* worlds. By implication, a state of dependency emerges as continuous in human life, and the environment therefore as continuously important.
> *(Ellsworth, 2005, p. xii)*

Winnicott's predecessors, Freud and Klein, had overlooked the continuous relation between humans-nature, body-environment, and subject-object as interpenetrating and mutually conditioning dependencies. Contemporary theory in early childhood education and childhood studies has only recently considered the vitality affects and animacies of children's object-relations (c.f. MacRae, 2019), while social cultural theory (Brooker et al., 2014) and cultural-historical theory (Colliver and Fleer, 2016; Fleer, 2010) continue to dominate minority Western thought on child development and learning. Winnicott's (1989) work is thus considered valuable in retheorising the animate environmentality of object-relations in studies of childhood play and education that break with cognitivist and rationalist traditions (c.f. Bibby, 2017). However, such work also reveals the need for critical revisions to Winnicott's work in light of post-developmental theories of childhood. As Winnicott writes in the tailpiece of *Playing and Reality*:

> I am proposing that there is a stage in the development of human beings that comes before objectivity and perceptibility. At the theoretical beginning a baby can be said to live in a subjective or conceptual world. The change from the primary state to one in which objective perception is possible is not only a matter of inherent or inherited growth process; it needs in addition an environmental minimum. It belongs to the whole vast theme of the individual travelling from dependence towards independence.
>
> *(Winnicott, 1989, p. 204)*

Here Winnicott argues that all humans begin their lives immersed in a subjective and conceptual world, with object-relating always occurring in oscillation with subjective experience as children move towards increased independence and capacitation, gradually letting the chaos of the alien outside come in, and learning to modulate the fluxes of affectivity and vitality. Yet, as many eco-feminist and posthumanist theorists of childhood have made clear (Lee, 2001; Murris, 2016a; Taylor et al., 2012), the movement from dependency to independence is dangerously overemphasised in conventional accounts of child development, which privilege the sovereignty of the bounded humanist subject as master of both self and environment. As Taylor, Paccini-Ketchabaw, and Blaise argue,

> The notion of the autonomous individual child perpetuated by child development theory is not only an illusion, it is also a grossly inadequate conceptual framework for responding to the challenges of growing up in an increasingly complex, mixed-up, boundary blurring, heterogeneous, interdependent and ethically confronting world.
>
> *(Taylor et al., 2012, p. 81)*

When the complexity and heterogeneity of worldly life are taken into account, the pathway from childhood to adulthood can never be predetermined according to a normative developmental model. Similarly, it becomes clear that there is never a stage or moment of human life that transcends or masters the socioecological dependencies which are the relational conditions for existence. Rather than achieving independence, there is an inescapable movement from dependencies to dependencies. This means that the movement of subjectivity from child to adulthood is never divorced from the dynamic evolution of the environment as a system of mutual dependency relations. The question for child-framed research, in this regard, is no longer how to cultivate children as independent researchers, learners, or social agents. Instead, the question is how the *quality* and *nature* of dependency relations shift when children enter into dynamically changing constellations of events and encounters.

Winnicott's (1989) relational and dynamic notion of the playspace as an eco-developmental system suggests that all learning is predicated on transitions *through* the environment as a meshwork of relations that forms the very condition for life. Thinking alongside post-developmental theories of childhood, our intervention into this theory of play involves questioning Winnicott's overarching narrative of "dependence to independence" as a teleological progression. Indeed, from the posthumanist perspective that we cultivate throughout this book, there is no possibility of independence from worldly entanglements. There is only the ongoing movement from attachment to attachment, dependency to dependency. This resonates with Barad's (2007) agential realism which poses an inherent indeterminacy and intra-dependency of events, as well as Whitehead's speculative philosophy which we discuss in detail in Chapter 5. There are no "independent agents" in these posthumanist ontologies, just as there are no independent agents in many Indigenous cosmologies founded on the inseparability of subjectivity from nature (c.f. Bawaka Country et al., 2016; Rosiek et al., 2020).

Anomalous Pedagogies – *Opening Up Transitional Spaces*

In addition to theories of play, our assembling of the "research playspace" as a methodological figuration was informed by theories of affective pedagogy developed in fields such as cultural studies (Hickey-Moody, 2013) and contemporary art (Ellsworth, 2005). In *Places of Learning*, Ellsworth (2005) draws on Massumi's theory of affect as pre-individual intensity to study "anomalous pedagogies" of difference, uncertainty, disjunction, noncompliance, and alterity that emerge through works of contemporary art, architecture, media, and design. For Ellsworth, pedagogy is nothing less than the "orchestration of forces, sensations, stories, invitations, habits, media, time, space, ideas, language, objects, images, and sounds intended, precisely, to move the materiality of our minds/brains and bodies into relation with the other material elements of our world" (p. 24). These myriad elements are brought together to constitute anomalous places of learning that operate as pedagogical pivot points that reconfigure the dynamics between intensive and extensive milieus. Ellsworth describes pedagogies as anomalous when they are "peculiar, irregular, abnormal and difficult to classify", moving in elliptical orbits that swing far out from ordinary educational discourse (p. 5). Such pedagogies are embedded within the experiences of aesthetic places of learning such as architectural spaces, art installations, relational events, and interactive media platforms. Anomalous pedagogies are defined by their capacity to disrupt and reconfigure the relations between inside and outside, public and private, and the ecological and aesthetic dimensions of experience through affective experimentation. They do so by operating through a logic of feeling that redistributes the possibilities for affective participation, transition, and engagement. As Ellsworth explains:

> Pedagogy, like painting, sculpture, or music, can be magical in its artful manipulation of inner ways of knowing into a mutually transforming relationship with outer events, selves, objects, and ideas. Or, it can used to simply manipulate, through congealed forms, unresponsive shapes, and derivative logics.
>
> *(Ellsworth, 2005, p. 7)*

Pedagogy becomes a curatorial process that sets up the possibility conditions for the emergence of new affective ecologies of movement, sensation, participation and belonging (Hickey-Moody, 2013; Rousell, 2021). As noted in the previous section, we don't see how these affective pedagogical ecologies can be reduced to the work of "independent agents", such as individual teachers, artists, or researchers. Moreover, as Manning notes, these learning ecologies are always already more-than-human, expressing as they do "the force of becoming that is singularly attendant to an ecology in the making, an ecology that can never be subsumed to the artist or to the individual participant" (Manning, 2016, p. 63). This also means that the outcomes of anomalous pedagogies can never be specified or even knowable in advance, but must be left open, pragmatically, to the relational play of affectivity and movement in the emergence of the ecology itself. This became evident, in a range of ways, throughout the Climate Change and Me project, as children developed experimental practices of environmental film-making, photography, performance, and creative writing that consistently diverged from any predetermined aims or expectations for what might ensue.

This openness to difference and noncompliance with expectation is crucial to understanding pedagogy as an ecological and aesthetic process within the research playspace. Rather than being the art or philosophy of a teacher who teaches, pedagogy becomes the artfulness of an environment that activates the tendencies for learning to circulate through multiple bodies and selves in the making. This reflects a concern with "what pedagogy *does*, rather than what it means or how it means" (Ellsworth, 2005, p. 27). What pedagogy can do, in this new materialist sense, is a question of how architectures of engagement are collectively designed, encountered, activated, and mobilised "in ways that modulate intensity, rhythm, passage through space, duration through time, aesthetic experience, and spatial expansion and compression" (p. 42).

Learning Collective Self – *Becoming Collective Space*

In taking up Ellsworth's (2005) concept of pedagogy as an affective, material, experimental, and emergent event, this inquiry distances itself from certain critical, social constructivist, sociocultural, and phenomenological orientations towards learning that rely on a cohesive individual subject and its representations as the units of analysis. As Ellsworth (2005) argues, "such notions are underwritten by assumptions that there is an identifiable self, a locatable point of view or subject position from which meanings are made and through which experience is organised and held together" (p. 7). Akin to what Whitehead (1978) critiqued as the "fallacy of simply location", such figurations of a bounded and locatable subjectivity capture the experience of learning within an all-too-human image of thought, such that "the personal is prefigured or 'prereflected' in the world, in a closed loop of 'intentionality'" (Massumi, 2002, p. 191).

In proposing an alternative account of learning grounded in theories of affect, Ellsworth (2005, p. 5) develops the concept of the "learning self" to describe the collective experience of learning as a material process of becoming. This means that the learning self is never attributed to a fixed or static body but, rather, to *a body in motion* that is open to "the possibility of qualitative transformation- of thinking and being something unprecedented" (p. 121). Movement is a primary element of the learning self as a collective process of becoming that is simultaneously physical, conceptual, sensory, and affective. This learning process is noncompliant not only with what might be demanded of it; it is also noncompliant with its own prior patterns of expression and instantiation. As Manning writes,

> A practice of collective learning is about the movement of thought, engaging thought at the immanent limit where it is still fully in the act. Learning happens through us, with us. We are bearers of thought in the sense that it is carried along.
>
> *(Manning, 2019, p. 47)*

This focus on thought as a variation of *movement* allows the concept of the learning self to situate the practice of learning within the embodied experience of "change taking place", as a zone of indeterminacy that "smudges" the boundaries between body and environment, inside and outside, self and other (Ellsworth, 2005, p. 122). This involves a process of learning in which the body "is blurred by its own indeterminacy and by its openness to an elsewhere and to an otherwise" (p. 122). At the practical level, this challenges received notions of "embodied" learning in environmental education which simply take the body as a phenomenological container or "unit" of experience which learns through physical interactions with the environment. We discuss the practical implications of this "blurring" of the learning body as processually indistinct from its environment in Chapter 5, where we draw on examples of children's image-making practice which directly implicate their bodily movements in the environmental milieu (Figure 3.1).

Lee (2008) further draws out the implications of this distinctive conceptualisation of embodied learning through a close analysis of Ellsworth's work:

FIGURE 3.1 Wax transfer photographic image created by child researchers for the *Past Now Future* exhibitions series.

> Learning's basis in the body, which is essentially precognitive, is not necessarily what Ellsworth means by materiality or embodiment. Because the bodily aspect of learning occurs before cognition, the learning self is ultimately not tied to one particular body- embodiment has "ontological primacy over cognition".
>
> *(Lee, 2008, p. 197)*

This reading helps to further contextualise Ellsworth's concept of the learning self within Whitehead's process-orientated account of subjectivity. For Whitehead (1978, p. 29) the subject is "not an entity that projects towards the world, or that phenomenologically 'intends' the world, but rather one that is only born in the very course of its encounter with the world ('subject'), and that gets precipitated out of this encounter, like a salt precipitated out of a solution ('superject')" (Shaviro, 2009, p. 21). Here the self is precipitated in and through the event of learning as an encounter in which the actual world is sensed, felt, and encountered differently. The learning self thus constitutes a double articulation of becoming that is characterised by its potential to vary from itself, where subjectivity is "becoming-other" while also perpetuating its own existence (Deleuze and Guattari, 1994, p. 177).

As Ellsworth (2005, p. 35) notes, the learning self that emerges from an encounter with the world is "at the same time unmistakable and unprecedented ... [it] never emerges in the same way twice. In its newness, it is comparable only to itself. It is singular, without model and without resemblance". Every instance of learning is a new event that also constitutes a new self. The new self that emerges is not the locus of a personal identity, but a distributed and collective body in motion that is entangled with legions of other bodies. Manning (2013, p. 127) writes that "there is no body that isn't always already collective, always already active in the relational interweaving of more than one ... ecology in the making". A body, for Manning, is *always more than one*. As a body in movement, the learning self is a collective individuation of subjectivity that composes itself "in the same way that an artist creates new forms from the palette" (Guattari, 1995, p. 7). In this sense, the body is learning to construct a collective self, and indeed, a collective space where new modes of subjectivity can be experimented.

Trans-Qualitative Research Playspaces

A posthuman research playspace is a collective learning event. It operates not through a logic of extraction or identity, but through a logic of feeling (Guattari, 2008) that enacts micropolitical and emergent relations of care for the collective process of inquiry as it comes into mutual expression. Play is the way. In this respect, the playspace is not yoked to conventions of academic knowledge production or outputs that demand a predetermined form but, rather, takes shape through "para-academic" practices of critical and creative experimentation. What happens in the research playspace cannot be quantified, calculated, measured, categorised, coded, or even, strictly speaking, defined. It eludes language, interpretation, and explanation. Its practices draw more so from the domains of activism and the arts than from the normative conventions of the social sciences. The research playspace offers a radical proposition for what inquiry might become in the "afterward" of conventional qualitative methodology (c.f. Lather and St Pierre, 2013; MacLure, 2013; Rousell, 2019; St Pierre, 2011, 2017, 2019).

To the extent that the research playspace generates condition for producing subjectivity and aesthetic experience beyond the limitations of the liberal humanist subject, it is precisely the ecological character and speculative potentials of the "qualitative" that is at stake in our experimental work with children and young people. We have described such approaches as "*trans-qualitative*" rather than "post-qualitative" in recent work (Rousell, 2021; Rousell and Cutter-Mackenzie-Knowles, 2020b), as we are not trying to move beyond or give up on the qualitative but, rather, to foreground the radically qualitative dynamics of all events and experiences. As Rousell (2021) explains,

> the notion of 'trans-qualitative inquiry' invokes a resistance to abandon the primacy of the qualitative in sensing the intensities of felt experience, particularly when concepts of 'feeling', 'sense', and 'experience' are no longer limited or modelled on human subjects, selves, or individuals.
>
> *(p. xxix)*

What techniques might allow us to collectively craft and inhabit posthuman research playspaces within the highly regulated "humanising" environments of public schools, libraries, and universities?

Crafting the Research Playspace

The Climate Change + Me project offered a rare testing ground for developing methods of working and thinking together as an intergenerational research collective responding to climate change as a complex, multidimensional problematic. The project was funded by the NSW (New South Wales, Australia) Environmental Trust, which is an independent statutory body established by the NSW government to fund a broad range of organizations to undertake projects that enhance the environment of NSW. Working towards the NSW 2021 goals, the Climate Change and Me project aimed to strengthen local environments and communities by increasing opportunities for children and young people to be proactive in climate change education research. The core audience of the Climate Change and Me program were children and young people aged 9–14 in Northern regional NSW. This audience was selected because children are often targeted but rarely consulted as legitimate contributors to educational research and curriculum development (Kellett, 2005).

The overarching research aim of the *Climate Change and Me* project was to map the diversity of young peoples' shifting awareness, attitudes, and actions towards climate change through a child-framed research methodology. The methodology aimed to establish research playspaces as platforms for young people to develop their own creative approaches to mapping their shifting experiences and understandings of climate change. This creative, socially engaged, and action-driven process of co-production then led to the co-development of an online social media platform and network; the *Past Now Future* community exhibitions in public spaces; transdisciplinary, co-designed Climate Change Curriculum; and the Climate Change Challenge, a community event which brought together local schools, climate scientists, environmental artists and writers, and members of the wider community to address the challenges of climate change. Through this emergent process of co-production, we deliberately allowed for the creative practices and theorizations of young people to dynamically impact on the methodology itself as it was unfolding in real time. This creative methodology required us to develop alternatives ways of approaching and working with school communities as co-creators of posthuman research playspaces.

The Climate Change + Me project allowed for previous frameworks of child-framed research to be extended into the areas of research-creation (Manning and Massumi, 2014; Rousell, 2017, 2020) as well as post-qualitative (Lather and St Pierre, 2013) and trans-qualitative inquiry (Rousell and Cutter-Mackenzie-Knowles, 2020b). While the children were initially introduced to techniques associated with visual ethnography, such as photography and video practices, we also supported them in extending their inquiries through creative practices of their own choosing. This included providing children with specialized art materials based on the directions their lines of inquiry were taking, as well as opening lines of collaboration between children from different schools. For example, one child researcher began to construct a speculative fictional world through detailed drawings of characters and textual entries on the project website. We mailed her a package of specialized drawing materials to support this work, while also providing avenues to extend her work into a collaborative writing and art experiment on the project website. We describe the development of this particular strand of work in Chapter 6. These extensions of the methodology through children's poetry, speculative fiction, drawing, and dramatic works were then brought back into the workshops to stimulate collective responses from the wider research cohort. In this way, the children were supported in developing their own approaches to the project that often exceeded or broke with our expectations, and these emergent, child-driven practices were then allowed to impact on the overall methodological trajectory of the project itself.

As we discussed in Chapter 1, the project took place within a highly situated socioecological context in the subtropical North of New South Wales on the unceded territories of the Bundjalung Aboriginal Nation. The region is divided into local government areas known as "shires", each with their own councils and chambers of commerce that differ in their political commitments, socioecological orientations, and development agendas. While the coastal areas surrounding Byron Bay are known for being international tourism, retirement, and alternative lifestyle destinations, the rural inland areas remain largely reliant on high-impact agricultural industries such as beef and dairy farming. The near-absence of public transportation means that these different areas across the region are typically isolated from each other, and dependent on private cars to access other towns and villages. As a result there can be very significant differences in cultural contexts and qualities of life across the region. We wanted to be inclusive of these differentials as we began canvasing schools across the region in the early stages of the project.

As we began the project in early 2014, climate change was not nearly as publicly visible or widely accepted an issue as it is now, in 2021, as we complete this book. The certainty and exactitude of climate science have advanced significantly in the past seven years, as have social engagement and action on the issue, in no small part due to the 2019 international youth climate strikes. Because the age range for project participants was between 9 and 14, we were tasked with recruiting both primary (years 4–6) and secondary (years 7–8) schools into the project. We initially contacted every primary and secondary school in the region by phone and email with details about the project, and an invitation for students to work with us on the project as researchers. Interestingly, while Byron Shire has some of the highest median house prices in Australia and is widely known for its alternative "green" lifestyle culture, not a single primary or secondary school from Byron Shire agreed to participate in the research stage of the project. Surprisingly, the schools that did express interest were from shires such as Ballina, Tweed, and Kyogle with significant populations of working class and agricultural communities and substantially lower median incomes and costs of living. This initial finding upended any assumption that schools in wealthy, privileged, "green", or "alternative" communities would be first takers on a project that promised to give children and young people a platform to research and address climate change on their own terms. It also meant that our research playspaces were constructed with children and young people in rural and regional communities that were not generally orientated towards green politics or environmental justice initiatives.

After several months of recruitment efforts we ended up with three primary schools and two high schools that committed to the project from the Tweed, Kyogle, and Ballina Shires. In each case, the school's interest in our project was galvanised by an exemplary teacher who recognised the urgency and value of the project's aims and approach, as well as the potential benefits the project might have for their student cohorts. Teachers initially distributed information to their students about the project, and we also visited schools in person to discuss the contours of the project in more detail with prospective student researchers. The project was oversubscribed with young people as we were only funded to involve 100 child researchers, but we made the ethical decision not to exclude any children who wanted to participate. In the end we had 135 children and young people join the project as active researchers. Together with these children and young people we formed an intergenerational collective, a community of inquiry.

The initial activation platforms for this community of inquiry consisted of full-day workshops designed to mobilise children's research into climate change within their own communities and environments. Our collective coming-together commenced with an initial series of activities specifically focusing on the Anthropocene concept as a propositional entry point into climate change education research. The young researchers were not given predetermined definitions or perspectives on the Anthropocene and climate change, but were encouraged to develop their own approaches to researching these concepts and phenomena in relation to their own lives, including through the use of interviews, video, photography, field notes, drawings, and poetry. As advocated by Barratt et al., (2013), the research workshops commenced with a series of opening activities focussed on drawing out young people's pre-conceptions (their own and others) of climate change before presenting the actual proposition for them to develop their own research trajectories. These opening activities included an interactive snowball, scrum, and Post-it mapping exercises.

Following these opening activities, the focus turned to the concept of the Anthropocene in contextualising climate change as a human impact at the geologic scale of the Earth's planetary epochs. The young researchers had not come across the concept of Anthropocene previously, and were enamoured with its resonances of "deep time". We related this concept to the disciplinary field of anthropology which children and young people were more familiar with, opening a discussion on the study of the relationship between "Anthropos" or "human" and the Earth at planetary scale (Figure 3.2). We also considered the problematic coloniality of conceptualisations of the Anthropocene and anthropology, and their disconnect with Indigenous cultures who have lived under the threat of extinction since colonisation (Whyte, 2018). These deep geophilosophical discussions led to a collectively generated series of propositions for studying the shifting relationships between young people and the Earth as a whole. When we then presented young people with the proposition to generate their own inquiries, we did so with the supposition that

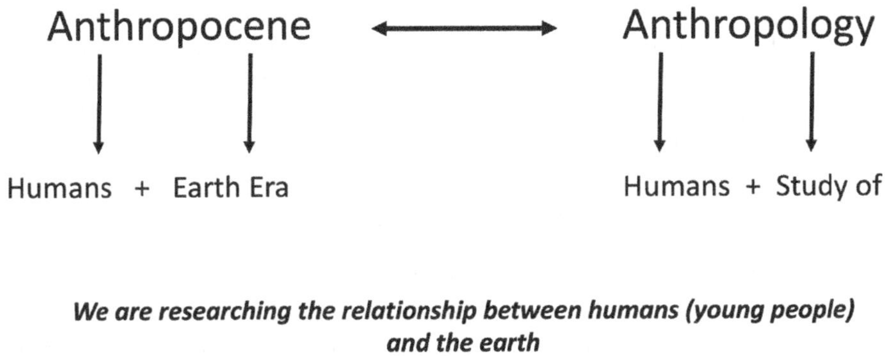

FIGURE 3.2 Workshop slide used to provoke critical propositional conversations that mobilised children and young people's research into climate change and Anthropocene in their communities.

these research trajectories would continually shift across the life of the project, and would inevitably come to change the aims and propositions of the project as a whole. These initial experiences were foundational in collectively crafting our Climate Change + Me research playspaces.

Breaking Away from Traditional Child-Framed Research Methods

A unique contribution of this research to the development of child-framed methodologies has been the construction of research playspaces in which children and young people develop the skills, experience, and support needed to cultivate lines of inquiry that break with adult expectations and parameters. This sensibility directly informed our workshop designs which aimed to afford the flexible stability children needed to take creative risks in the development of their own research practices. In this way, the children were supported in developing their own approaches to the project that often exceeded or broke with our own expectations, and these emergent, child- driven practices were then allowed to impact on the overall methodological trajectory of the project itself.

We are particularly interested in how "breakaway" lines of inquiry were able to emerge through posthuman research playspaces, and how these breakaways came to disrupt conventional modalities of research with children, including child-framed approaches that purport to put children at the centre of things. Such approaches have often referred, for example, to hierarchical "ladders" of participation which progress from extractive and manipulated participation (at the bottom), to assigned and informed participation (in the middle), to child-initiated research (at the top) (c.f. Hart, 1997). While we acknowledge the value of a child-initiated approach, we resist the reduction of such an approach to a hierarchical ladder which relies on essentialized categories of individual participation and initiation. Our aim in activating research playspaces was not to centre everything back onto the child as sole "initiator" of the research but, rather, to form an intergenerational community of inquiry where different roles and agencies could remain fluid, relational, and dynamically changing. Indeed, this is one of the fundamental differences between trans- or post-qualitative inquiry and conventional qualitative research, including many child-framed approaches, which presuppose a static and contained "unit of self" as subject and participant in research. Rather than starting from the assumption that subjectivity can be simply located or contained in the individual (c.f. Whitehead, 1978), trans- and post-qualitative approaches start with an immanent field of experience in which "all subjects are transsubjective and transindividual" (Massumi, 2018, p. 61, emphasis in original).

Within the research playspace, as in the theories of transindividual play and learning we elaborated in previous sections, it is not clear who (if anyone) is the initiator of a given event. Rather, it is the event itself as a relational fielding of experience that initiates and mobilises the inquiry (Manning, 2013; Rousell, 2021), not individual (adult or child) human researchers. The research playspace is therefore better understood as a propositional space of transindividual activation rather than a determining space of initiation. We can, in this sense, understand the research playspace as framed by the shifting figuration of the *event* rather than the child as bounded individual, where the "breakaway" trajectories of inquiry arise not from (adult or child) individuals, but through *breakaways of the event* as a runaway process of collective experimentation which has no locatable point of origin in one (child or adult) body or another.

Video as Transitional Medium

In our iterative workshop designs and approaches, we found that giving children time to explore creative methods without any adult intervention often led to these breakaway lines of inquiry. During the primary research stage of the project, children and young people used iPads as mobile recording devices to create their own responses to climate change in their schools, homes, and extended communities. A number of children chose to conduct interviews with their peers, parents, siblings, and teachers, many of which revealed a thick plurality of responses and understandings of climate change in the children's homes, neighbourhoods, and local communities. Other researchers were inspired to use the iPads in more provocative ways, resulting in a wide variety of video works that included interviews with trees, improvisational forest romps, upside-down political discussions, an imaginary tea party, and a future world populated by digital avatars and virtual environments.

One such example was where a young boy (10 years old, Daniel) elected to interview a tree about their perspectives on climate change. The transcript follows:

INTERVIEWER: So, what's your opinion on climate change?
TREE: 2-second pause
INTERVIEWER: Do you have any solutions for climate change? Okay next question.
TREE: 3-second pause

INTERVIEWER: How do you feel about climate change?
TREE: 3-second pause
INTERVIEWER: Stupid tree
INTERVIEWER: Do you think Climate Change is a threat?
TREE: 5-second pause
INTERVIEWER: Um, is climate change a good thing? Is it a bad thing?
TREE: 5-second pause
INTERVIEWER: I really don't have any other questions, other than do you have any worries about climate change?
TREE: 3-second pause
INTERVIEWER: No, well thank you. Thanks for your opinion.

Daniel's explanation of interviewing a tree was simple – "a tree is a being who stores carbon – why wouldn't I ask their opinion?" His response was somewhat tongue in cheek and playful, but at the same time the iPad provided Daniel with a creative impulse which Winnicott (1989) maintained occurs in learning when creative thought and experimentation place internal and external worlds into transformative relation (Cutter-Mackenzie-Knowles and Rousell, 2020). The interview offers an opening into the phenomena of climate change from the perspective of Daniel's encounter with the tree; specifically, the questions he was asking himself about the entanglement between himself, the tree, and climate change. Thinking with plants as agents and respondents within shared biocultural environments introduces practices that question the conventions and expectations of knowledge and learning. While in this case the engagement was relatively short, Daniel's conversation could easily have expanded into studies of tree communication through mycorrhizal fungal networks and airborne chemical exchanges, which perform a key role in supporting communities of trees and their offspring (Wohlleben, 2016). The virtual possibilities of Daniel's iPad chat with a fellow socioecological traveller breakaway from received knowledges, stirring up the possibilities for multi-species inquiry into nature as event (Debaise, 2017; Young and Cutter-Mackenzie-Knowles, 2020).

We encouraged this playful "misuse" of the iPads for creative purposes that broke the traditional conventions of qualitative research and, in some cases, abandoned common sense appeals to rationality, discourse, and sensemaking. As the research extended from the initial research workshops into homes and extended environments of children's lives, these odd video creations came to form a fascinating archive of creative artefacts that generated a rich pattern of audiovisual textures and political aesthetic sensibilities. We found that these videos often rendered what children might think, say, or do when no adults were present to supervise them or interfere with their ideas and activities. In this sense, the iPad came to operate as a transitional object that enabled children to venture into new territories of thinking and action, creating both a mirror (opacity) and a window (transparency) into children's worlds at the same time.

In recent work we have begun to theorise the role of digital video in the co-constitution of research playspaces with children and young people in terms of a "transitional medium" (Cutter-Mackenzie-Knowles and Rousell, 2020). We note how children invented ways of using iPads as film-making devices that mobilized the research playspace in particular ways, effectively folding the body of the child into the milieu of the surrounding environment. Given the large size of the iPad interface, children are able to see what is being filmed as they move dynamically with the environment, thus modulating the medium of video in ways that are performative, unpredictable, and impulsive. In this sense, the screen becomes both a recording surface and a surface of projection for activating transitional space through environmental play, bringing the "audience" of the film into direct contact with the process of co-production. We argue that this enables the film to be produced and viewed by children at the same time, adding a dimension of hyper-sensibility as well as a public intimacy that is specific to the medium being employed. Through the digital interface of the iPad's screen, the medium of video becomes transitional in its capacity to put children's creative impulses and dynamic social milieus into transformative relation with a shared environmental outside, as well as an audience that is immanent in the production of the film itself (Cutter-Mackenzie-Knowles and Rousell, 2019; Cutter-Mackenzie-Knowles and Rousell, 2020).

Another Breakaway – Film-Makers

In many ways, the experimental videos produced by children in the CC+Me project are more closely aligned with video art than the conventions of naturalistic observation and ethnographic documentary cinema. Children's experimental and playful usages of video as a transformative medium also break with the traditional conventions of video methodologies in qualitative educational research. As de Freitas (2016) notes, such conventions are tied to the history of scientific film and the vagaries of visual documentation as truthful and accurate representations of empirical phenomena. Rather than documenting and analysing children in films as cultural representations or scientific evidence, we became interested in how *children were actually producing films* by using video as a transitional medium that enables them to think and interact with climate change in novel ways. By focusing on children's co-productive engagement with video as a transitional medium, we acknowledge that *we still do not know what children can do*. What happens when we give children iPads and ask them to go out and make films exploring the ways that climate change is affecting their lives? What happens when we bring those films back in, analyse them together, and ask them to make different films in response to what other children have produced? How do these film-making practices elaborate a way of thinking ecologically about childhood, education, and climate change? These are the kinds of open-ended questions and experimental approaches that we pursued in the development of posthuman playspaces for climate change education research.

In the following vignettes, we discuss three examples of experimental video works produced by children which offer views into their research playspaces in the

Climate Change and Me project. These examples provide brief openings into children's environmental thinking, imagination, and play in response to the question of how climate change is affecting their lives, communities, and environments. The first focuses on children's sense of invisibility and a dehumanizing lack of agency in relation to climate change; the second foregrounds the improvisational, animalistic, and often irrational nature of children's "wild" environmental play; and the third addresses children's experimental use of digital video to generate a dystopian virtual world.

Becoming Invisible

This film (see Figure 3.3) emerged from a discussion between three child researchers in a year five class regarding their feelings of invisibility, disconnection, and lack of political agency in relation to climate change. While they felt that children's experiences and perceptions of climate change were not being valued or acknowledge by adults, they also noted a sense of complicity in children's contribution to climate change. They began to discuss children's lack of understanding and their desire to know more about climate change, as well the need for children to actively change their own habitual practices and patterns of thought. As the researchers were developing various ideas for creating a film that would embody and articulate these discussions, they found a collection of gold and silver masks in the storage room adjoining the classroom. They decided to use the masks to conceal their faces and identities in the film, emphasising the invisible and unknown qualities of children's voices. As they began to play with the possibilities of using the masks, they decided to create a series of written signs that could be used to articulate their questions about children's relationship to climate change. They devised a cascading series of questions, which began with the phrase "Climate Change and Children…", and then proceeded with: "Do they know? Do they care? How can we teach them? How can we change ourselves?" These open-ended questions ended up forming a discursive structure for a performative and affectively responsive series of actions which comprised the resulting film.

After inviting another student to serve as video operator, the three researchers arranged themselves in a line facing the camera with a garden area behind them. As the film progresses, they take turns moving into the front position of the line as each piece of text is revealed sequentially. This movement begins sporadically but slowly takes on a peculiar and often uncomfortable rhythm, as the stops, starts, and misshapen notes of the school band's practice begins to infect the performative event taking shape. In the resulting film, the researchers appear faceless and dehumanised while their bodies and their questions move in and out of synch with a "found" soundtrack. These disjunctive spaces between the sounds and images embedded in the film produce a surreal quality, as the initial proposition for the film as a social intervention becomes rhythmically warped by the sound field of everyday school events occurring beyond the frame of the camera.

FIGURE 3.3 Becoming invisible.

Wild Play

In this second example, another group of three child researchers take an iPad into a forested area of the school without any initial planning or preparation. The resulting film unfolds completely spontaneously, beginning with one researcher running madly through the forest and crashing into another, who ends up falling into the bushes and then scrambling on hands and knees along the ground. This initial scene of outright silliness sets the tone for a series of playful and often ridiculous interactions between the researchers and the surrounding environment. There is an intoxicating quality to this series of behaviours, as the silliness is pushed to increasing degrees of intensity over the course of two minutes. There is also a consistent

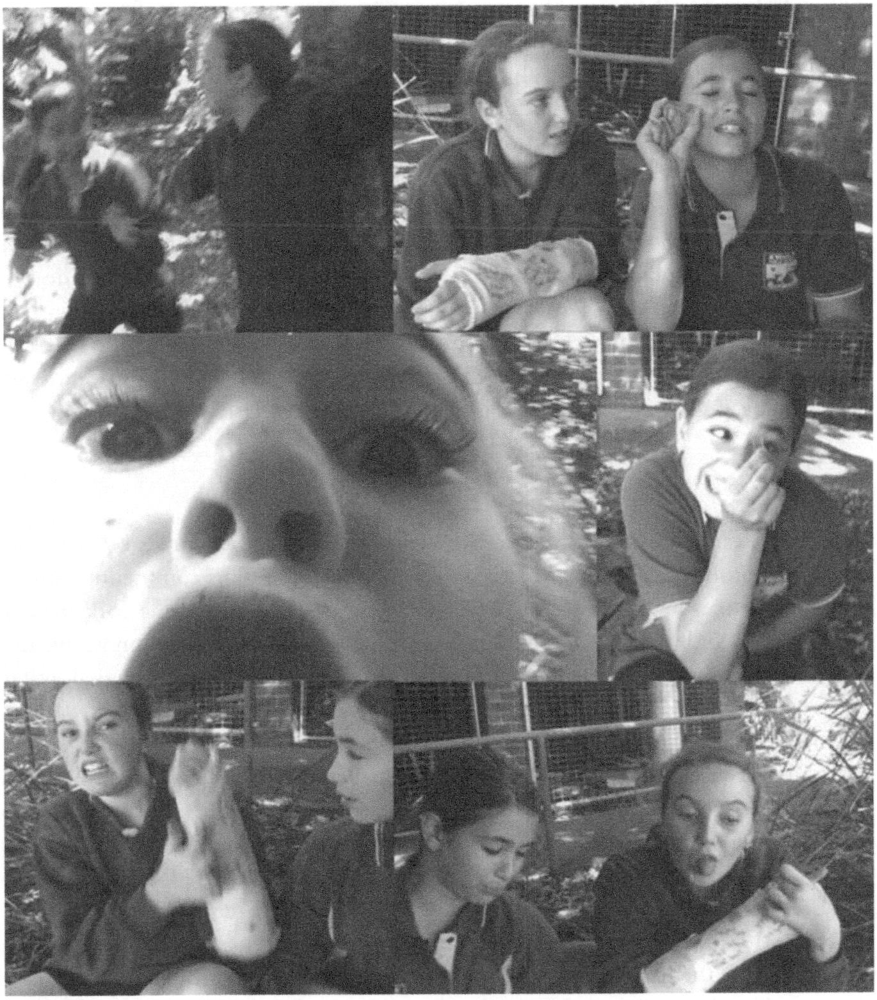

FIGURE 3.4 Wild play pivoting around the shard of glass.

sense of wildness and animalistic behaviour that infuses this film, as the researchers continuously use their bodies to explore their relationships with climate change in strange and unexpected ways.

Following the opening sequence of playful encounter and excitation, the researchers sit down on the ground and begin to develop a call-and-response chant, riffing off rhythmic variations of the phrase "what do ya think about climate change?" This chant is accompanied by a series of improvised dance moves, with each responding to the other in a call-and-response formation. At a certain point in these choreographic interactions, one researcher places her hand into a hole between the roots of a nearby tree. When she pulls her hand out, she is amazed to find that she has withdrawn a shard of glass that was hidden in this little hole. This piece of glass then becomes the impetus for a series of unusual performative gestures, including demonstrations of the shard's ability to leave visible scratch marks on the skin of one girl's arm. This is followed by another crude demonstration of what would happen if somebody sat on the shard of glass, complete with hand gestures suggesting the penetration of one object into another. Throughout this sequence, the researchers display a range of intense and dramatically charged facial expressions which are often addressed directly to the camera, as well as to each other. After some loose discussion of the shard's relevance to the hidden dangers of climate change and humanity's impact on the Earth's geologic strata, the film finishes with a spasmodic gesture in which one of the researchers throws the shard directly towards the camera, narrowly missing the camera operator whose face suddenly appears in stark surprise. At this point, the researchers collapse onto each other in compulsive fits of laughter and uncontrollable hilarity (see Figure 3.4).

After Earth

Our third example is a film produced by two child researchers in year 7, who had previously produced a number of insightful and engaging films commenting on the effects of climate change on such everyday practices as doing homework, playing video games, and making dinner. With this film, they developed an innovative idea that involved the creation of a dystopian future world comprising digital avatars and virtual environments. To produce this film, the researchers negotiated access to a nearby section of rainforest during lunchtime, an area that students were not normally allowed to visit. Working from a basic sketch of possible camera shots and action sequences, they proceeded to experiment iteratively with the affordances of the digital video format. The initial opening sequence of the film involved placing fingers over the camera lens of the iPad to produce an orange-red effect that gradually resolves into a scene: the motionless figure of a girl standing at the top of a staircase with a vacant expression, as shot from below. This figure becomes the silent protagonist of the film, a girl who realizes that she exists as a digital avatar of her former self and that she inhabits a virtual environment that is disintegrating around her. Her doll-like expressions bring the figuration to life without dialogue, as she appears unable to speak amidst the horrific discovery of her digitally encapsulated existence. By shooting into the afternoon sun, the film-makers used refractive light and lens flares to add a surreal quality to their shots. They also developed a special technique of wobbling the iPad in order to suggest the quivering failure of a virtual environment that is glitching out, and eventually shutting down.

One of the fascinating aspects of this film is the researchers' ability to produce the feeling of an artificial world and virtual existence without any spoken words, dialogue, or prompts. The story is told entirely by the physical movements and expressions of the female protagonist, who becomes not so much a character (in the conventional sense of a human subject) as a body that suddenly finds itself inhabiting an environment that is both toxic and strange. Her movements and facial expressions also suggest the birth or awakening of an artificial intelligence, perhaps a digital version of her original personality, whose initial interactions with the surrounding environment lead to a destabilisation and breakdown of the system.

Interestingly, we find the protagonist consistently peering directly into the camera's lens, as if searching the depths of the digital medium and indeed its implicit audience, for some sense of who (and where) she might be. The film concludes with her stricken expression of desperation as she falls to the ground and perishes, due to a virtual environment that is too unstable to support sentient life (see Figure 3.5).

FIGURE 3.5 Still shots from "After Earth".

Further Breakaways – Creative Analytic Thinktanks

These breakaway trajectories of researchers' film-making offer examples of creative practices activated by children during the initial research workshops, and then extended in their own time, both in and outside of school, for a one-year period. During this time, children also developed a range of other photographic, poetic, philosophical, speculative, and artistic practices as experimental research methods for mapping their shifting experiences and understandings of climate change. We were soon inundated with an extraordinary range of materials, artefacts, images, films, writings, poems, and other creative data. The Climate Change and Me website served as a key hub for intensive discussion and data sharing throughout this period, with children regularly posting their latest images, films, or writings on the site to share and discuss with other members of the intergenerational research community. Children often elected to post their data in the HangoutSpace we had created, where other children could engage with the data and open up extended discussions through the forum function. We carefully archived this material, both online and in physical files, and soon came to realise that the project had generated many thousands of photographs, and many hundreds of video files, drawings, and texts of various kinds. This deluge of child-generated data required us to develop specialised techniques for collectively analysing, composing, curating, and "making-sense" of an extraordinary range and intensive quantity of research material. We didn't want to simply walk away with all this data and interpret it ourselves. We needed to figure out techniques to enable the children to work with this data en masse in meaningful and impactful ways that could translate into public outcomes, including a travelling exhibition and co-designed climate change curriculum.

This led us to organize a series of analytic workshops called "ThinkTanks", which brought the intergenerational research collective back together to work with the data in schools and university spaces. Our approach involved splitting the data up into phonic types (images, films, and texts), and then designing specific techniques for collaborative analysis around each of these data types. In one Thinktank we brought several thousand images taken by children during the mapping phase, and spread them out on the floor of a school's gymnasium. The children then spent hours rifling through the photographs, selecting the most articulate images that "spoke" about climate change, and writing their own poetic annotations on the backs of the images. In subsequent workshops we converted a university science laboratory into an art studio to create encaustic wax transfers of these selected images for an installation in the Past Now Future exhibition. In another workshop we converted a high school lecture theatre into a cinema and screened a FilmFest of many of the hundreds of films produced by children during the project. Children sat in the audience and documented their thoughts and responses to each film, which led to the curation of interactive video installations for the Past Now Future exhibition. Other ThinkTank workshops focused on poetry writing, speculative fiction, drawing, and the editing of *The Changes*, a book of children's writing, art, and research which was published for the exhibition. By allowing our analytic process to become open and responsive to the creative possibilities of co-production posed by children themselves, the ThinkTank workshops effectively broke away from conventional analytic paradigms for child-framed and participatory research in education.

Past Now Future

One of the ways that our approach differs from previous studies of posthuman childhoods is our focus on children's climate change imaginaries as sites of aesthetic and ecological re-imagining of social, affective, and material worlds. Rather than teaching children *about* climate change and the environment through science lessons or outdoor learning activities, we continuously start with the question of *how they sense* climate change in the midst of their everyday aesthetic experiences, thoughts, feelings, lives, relationships, homes, communities, schools, media, neighbourhoods, cultures, places, and ecologies. This initial proposition inevitably provokes children to begin sensing, exploring, and researching their environments in a variety of different ways, including the environments of their own schools and community environments, cities, glaciers, reefs, and rainforests that they visited during school holidays, and even their own intimate family and home environments. This is, in Whitehead's (1967) terms, an invitation to enter into an *adventure* of aesthetic sensation and thought.

In equipping for this adventure, we introduced them to a series of tools and techniques for undertaking their own aesthetic inquiry as artists, philosophers, and researchers, including the use of iPads to conduct ethnographic interviews, take photographs, create videos, write essays, stories, and poems, and upload their research artefacts onto a shared online platform created especially for children and young people to use. This enabled us to follow the ways that children and young people were encountering climate change and the Anthropocene through their own experiential worlds, developing their own aesthetic frameworks and cultivating modes of inquiry that expressed their evolving aesthetic insights and understandings of climate change. As project leaders, our role was primarily that of creative facilitators and mentors for children and young people as they developed their own research and aesthetic sensibilities. We engaged in ongoing dialogues with them on the project website, supported their creative practices through the provision of materials and specialized mentorship, and continuously created online and physical spaces for co-creation, sharing, and analysis of the work that was being produced. This can be understood as a collective experiment in *aesthesis* in the pursuit of an unbifurcated account of children's aesthetic experiences of climate change.

A significant aspect of our approach was our commitment to work collectively with children and young people towards a shared goal of exhibiting their aesthetic experiences with the wider public. From the very first workshops, children and young people were aware that they were working towards a series of travelling exhibitions that would bring their aesthetic encounters and impressions of climate change to public attention. This gave many of the child researchers a sense of sustained purpose, or what Whitehead terms "appetition", driven by an urge to share and proliferate a collective aesthetic vision and imaginary across a wider field of public encounter. After an intensive process of brainstorming and discussion, the exhibition was named *Past Now Future* as suggested by Brent, a ten-year-old researcher from one of our participating primary schools. For Brent and his peers, the name *Past Now Future* encapsulated their aesthetic encounters with climate change and the Anthropocene as complex experiential events occurring across deep timescales, ranging from the geologic to the anthropological, to the immediate, the futural, speculative, and imaginary.

The *Past Now Future* exhibition took place over a three-month period in 2015, and was presented at eight public libraries in communities across Northern NSW. The exhibition was viewed by over 10,000 members of the public, and was also documented by local newspapers and on ABC National Radio through interviews with several of the young researchers. The CC+Me researchers were actively involved with all aspects of the exhibition as a curatorial and analytic process, including the titling, selection of works, locations, and artist/researcher statements. Groups of children worked together to compose these artist statements and invitations to the public, and presented their own public speeches to launch each of the exhibition events.

"Our goal with this exhibition is to show you what climate change looks like, what it sounds like, what it feels like, and what it might become in the future. Past Now Future is about us taking a stand and having a voice about climate change. It's about us researching the real life experiences and opinions of children and young people. It's about researching ourselves too, as kids who care about climate change and want to do something."

The *Past Now Future* exhibitions featured three major collaborative works. The first work, *Pieces of the Earth*, consisted of 135 digital images transferred into encaustic wax medium on plywood tiles (Figure 3.6), and is discussed in detail in Chapter 5. The second work, *Voices from the Anthropocene*, consisted of three hyperlinked video artworks exploring the aesthetic sensibilities, awareness, and actions of children and their communities in response to climate change (Figure 3.7). The third work takes the form of an edited exhibition book called *The Changes*, which features writings, poems, photo essays, drawings, and speculative fiction stories created by the philosophers/researchers (Figure 3.8), as discussed in Chapter 6.

Many of the writings in this exhibition book directly challenge both the ethics and aesthetics of normative, anthropocentric, and neoliberal life practices that contribute to the escalation of climate change. As 12-year-old researcher Nikki writes in her contribution to the collection:

> The difference must begin with us. We must make drastic changes to how we think and make choices about our everyday activities and focuses. We were not placed on this Earth to make an acquisitive and ideal life that supports the human race only; we were placed on this planet to foresee a life for all things, whether they are living, or non-living, plant or animal. Not to destroy the climate by means of our own self-centred needs.
>
> *(Nikki, 12-year-old researcher)*

FIGURE 3.6 Children creating photographic wax transfers and an interactive installation of their work in libraries and gallery spaces.

Breakaway Curriculum as Collective Inquiry Process

The final "breakaway" that we discuss here is a breakaway curriculum that emerged from the open-ended and emergent inquiry process that we describe above. This again constituted a significant break from previous models of participatory curriculum development, which have largely been consultative rather than genuinely creative and co-created with children as authors. Curriculum and curriculum writing are typically considered the work of teachers and educational policymakers. While there has been substantial progress in recognising the capabilities of young people in research, curriculum and policy have made little progress in this regard.

As part of the CC+Me grant (funding) application we emphatically stated that one of the major objectives of the project would be:

> The co-design and implementation of a child-framed climate change primary and secondary school curriculum and support resources that are locally and regionally relevant supporting ongoing environmental behaviour change and improvement in climate change.

FIGURE 3.7 Children engaging with the interactive "Voices from the Anthropocene" film installation in their local public library.

FIGURE 3.8 Selection of photos and drawings by children from *The Changes* exhibition book.

We were inspired by Kagawa and Selby's (2010, p. 242) argument that "climate change is too urgent and important to suffer 'death by formal curriculum'... calling for emergent curriculum approaches that embed climate change learning and action within community contexts". All too often climate change education is something "done" to children rather than led by children. Co-designing curriculum with young people was somewhat unknown territory to us as researchers and teachers. As experienced teachers we were well adapted in the practice of inquiry, but not writing curriculum with young people. Where to begin? We began by inviting CC+Me child researchers, their teachers and parents to expand our collective CC+Me research playspace to a curriculum playspace. Five teachers, two parents, and three child researchers joined our curriculum writing collective.

We embraced the research playspace methodology described earlier in the chapter as the entry point for curriculum making. Our initial workshops focussed on a "revisiting" of the research as analysed, curated, and publicly presented through the *Past Now Future* exhibition. This meant that the underlying content, processes, materials, and resources for the curriculum were all drawn directly from the work of children as researchers over previous phases of the project. Rather than asking children to "consult" and "have a say" on curriculum aims and materials prearticulated by adults, these workshops brought children, teachers, and researchers together to create a curriculum using materials *previously developed by children.* We did this by focusing on the collective inquiry processes that children had previously modelled in the research project, including the ways that they negotiated their own collective inquiry into climate change in their communities, and also developed their own creative methods and techniques for doing this work (see Chapter 3). In this way, the unpredictable and often disruptive process of "children-becoming-researchers" within the research playspace also came to comprise the underpinning machinery for the curriculum making process. In addition, the films, images, essays, interviews, poetry, and speculative fiction created by children in the project became the content or "textbook" examples to support the delivery of the curriculum in schools. In the end, we created a curriculum for students aged 9–14 that almost entirely comprised ideas, materials, and processes of inquiry devised by children through the research playspace.

Through an intensive process of cross-curricular mapping, our working group then managed to reverse engineer the phases of the inquiry process to meet the government-mandated outcomes for the key learning areas of Visual Arts, English, Humanities, and Science in the Australian National Curriculum. The curriculum proceeds through seven phases of inquiry-based learning built around processes that the child researchers had developed in the Climate Change and Me research:

Phase 1: What is Climate Change?

Introduce the unit of work as a collective inquiry into climate change and begin a research journal. Explore students' personal ideas, knowledge, and experiences of climate change. Learn about diverse cultural and scientific perspectives on climate change, including Indigenous perspectives, as well as its social and environmental effects.

Phase 2: Connecting with Climate Change

Relating climate change to the everyday lives of students. Students bring in an object, household product, story, song, video, photograph, news article, or other artefact that speaks to them about climate change. Students analyse the artefacts in groups and record in their journals.

Phase 3: Imagining Climate Change

Students engage with the complexities of climate change through their imaginations. Explore different climate change scenarios and future possibilities. Students create their own stories, characters, poems, or future imaginings in response to the CC + Me support resources provided.

Phase 4: Becoming a Researcher

Explore different ways of doing climate change research. Experiment with research methods such as online research, surveys, interviews, essays, fiction, video, and photography. Use examples of CC + Me student research in the support resources as starting points.

Phase 5: Working with Your Research

Bring together all the research your collective has done about Climate Change. Consider different meanings and themes, and how you might put your research together into an exhibition, performance, book, or other public presentation.

Phase 6: Communicating Your Research

Plan and develop engaging and sustainable ways to present your research to your communities.

Phase 7: Reflecting on Your Process

Evaluate and reflect on the collective inquiry process and what students and teachers have learned.

Each stage of the curriculum was flexibly structured to address standard learning outcomes in the disciplinary areas of Science, English, Creative Arts, and HSIE/ Geography. We released two versions of the Climate Change and Me curriculum online through the project website in 2016: one version for stages 2/3 (primary school) and one version for stages 3/4 (high school). The curriculum was accompanied by a text-based resource booklet containing examples of children's art, writing, images, poetry, and fiction from the original CC + Me project, as well as PowerPoint resources to support each phase of the collective inquiry process. We initially trialled the curriculum in six schools across New South Wales, which allowed us to evaluate the curriculum through responses from teachers and students. In her evaluation of the curriculum, one teacher from a local primary school describes how "the project achieved a high standard of curriculum that encourages students to become researchers, develop critical and ethical thinking, use imagination and personal expression, engage in higher order questioning, discussion and debate and become responsible global citizens". Other comments from teachers variously emphasized how the curriculum contains "comprehensive information about the history, present day and future of the world in terms of the effects of climate change on our planet", while also inviting students "to be part of a community-shared learning experience" and "understand the empowerment that children can have on important global issues".

Since the initial trial in 2016 we have had over 40 schools adopt the curriculum from across Australia, with teachers recognizing the interdisciplinary innovation and timeliness of a curriculum co-designed and developed with children working as philosophers and researchers of climate change education. We regularly receive emails from new teachers who have stumbled across the curriculum online, or through word of mouth, and found the child-led design and accompanying resources incredibly refreshing and enlivening for their students. We continue to embed this curriculum in our teacher education programs as an example of renegade curriculum production with children from the bottom up. As governmental curriculum initiatives continue to fail to effectively integrate and engage climate change studies in primary and secondary schools, the Climate Change and Me curriculum offers a micropolitical counter-narrative that has, in an increasing number of cases, effectively smuggled transdisciplinary climate change research into the formal school curriculum.

Conclusion

This chapter has outlined the contours of an emergent methodological approach to research with children through the co-construction and activation of research playspaces within intergenerational research collectives. Drawing on theories of environmental play, transitional space, and affective pedagogies that decentre the human(ist) individual, we have theorised the research playspace as a mobile zone of activation where there is no clear line of origin or initialisation. We have argued that child and adult researchers participate in the activation of research playspaces in ways that can't be determined in advance, where social agencies and roles are fluidified by the movements of collective events taking shape. To demonstrate how open-ended encounters in the research playspace can mobilise new trajectories of inquiry, we have described a series of "breakaways" where children led the Climate Change and Me project into new territory and actively generated new modes of inquiry which do not fit any existing methodological model or paradigm. Finally, we have described the co-production of a climate change curriculum based on the emergent inquiry processes of children within the research playpace, and designed precisely to inject the radical mobilisations of such spaces into the formalised systems of primary and secondary schooling. Rather than training children to think and work like conventional "academic" researchers, we have proposed research playspaces that foster children's work as researchers who actively disrupt the normative conventions of academic practice and the standardisation of learning and knowledge.

In navigating these propositions and breakaways through the research playspace, children learn to hold space with the work of artists, film-makers, performers, musicians, designers, and other critical and creative practitioners who generate collective thought experiments and develop alternative ways of thinking and making the world. In this sense, we see children becoming artists, philosophers, and researchers when they refuse to comply with a normative research protocol and associated adult expectations, and instead invent their own modes of co-creation and critical inquiry. Our work demonstrates how an open-ended, child-framed approach can lead to direct transformations of curriculum and a disruption of social hierarchies and stratifications that dominate conventional schooling. When children break with adult expectations, they show us something that we would never have been able to experience or understand otherwise. It is in those moments of noncompliance that children provoke us to think and do things differently, reconfiguring methodological concepts and practices in response to the critical crises and transformations of our times.

References

Barad, K. (2007). *Meeting the universe halfway: Quantum physics and the entanglement of matter and meaning*. Duke University Press.

Barratt Hacking, E., Cutter-Mackenzie, A., & Barrratt, R. (2013). Children as active researchers: The potential of environmental education research involving children. In R. Stevenson, A. Wals, M. Brody, & J. Dillon (Eds.), *The handbook of research on environmental education* (pp. 438–458). American Educational Research Association.

Bawaka Country, Wright, S., Suchet-Pearson, S., Lloyd, K., Burarrwanga, L., Ganambarr, R., … & Sweeney, J. (2016). Co-becoming Bawaka: Towards a relational understanding of place/space. *Progress in Human Geography*, *40*(4), 455–475.

Bell, N. (2008). Ethics in child research: Rights, reason and responsibilities. *Children's Geographies*, *6*(1), 7–20.

Bibby, T. (2017). *The creative self: Psychoanalysis, teaching and learning in the classroom*. Routledge.

Brooker, L., Blaise, M., & Edwards, S. (2014). *International handbook on play and learning*. SAGE Publishers.

Christensen, P., & James, A. (Eds.). (2000). *Research with children: Perspectives and practices*. Falmer Press.

Colliver, Y., & Fleer, M. (2016). 'I already know what I learned': Young children's perspectives on learning through play. *Early Child Development and Care, 1–12*. https://doi.org/10.1080/03004430.2015.1111880.

Cook-Sather, A. (2006). Sound, presence, and power: "Student voice" in educational research and reform. *Curriculum Inquiry*, *36*, 359–390.

Cook-Sather, A. (2007). Resisting the impositional potential of student voice work: Lessons for liberatory educational research from poststructuralist feminist critiques of critical pedagogy. *Discourse: Studies in the Cultural Politics of Education*, *28*(3), 389–403.

Cutter-Mackenzie-Knowles, A., Malone, K., & Barratt Hacking, E. (Eds.). (2020). *Research handbook on childhoodnature: Assemblages of childhood and nature research* (Springer International Handbooks of Education). Springer Nature.

Cutter-Mackenzie-Knowles, A., & Rousell, D. (2020). The mesh of playing, theorising and researching in the reality of climate change: Children's chldhoodnature research playspaces. In A. Cutter-Mackenzie-Knowles, K. Malone, & E. Barratt Hacking (Eds.), *Research handbook on childhoodnature: Assemblages of childhood and nature research* (pp. 199–222). Springer Nature.

de Freitas, E. (2016). The new empiricism of the fractal fold: Rethinking monadology in digital times. *Cultural Studies ↔ Critical Methodologies, 16*(2), 224–234, https://doi.org/10.1177/1532708616634733.

Debaise, D. (2017). *Nature as event: The lure of the possible*. Duke University Press.

Deleuze, G., & Guattari, F. (1994). *What is philosophy?* Columbia University Press.

Dewey, J. (1938). *Experience and education*. Kappa Delta Pi.

Dewey, J. (1956). *The child and the curriculum, and the school and society*. Chicago University Press.

Dodd-Nufrio, A. (2011). Reggio Emilia, Maria Montessori, and John Dewey: Dispelling teachers' misconceptions and understanding theoretical foundations (Editorial). *Early Childhood Education Journal, 39*(4), 235–237.

Ellsworth, E. (2005). *Places of learning: Media, architecture, Pedagogy*. Routledge.

Fielding, M. (2004). Transformative approaches to student voice: Theoretical underpinnings, recalcitrant realities. *British Educational Research Journal, 30*(2), 295–311.

Fleer, M. (2010). *Concepts in play: A cultural historical view of early learning and development*. Cambridge University Press.

Froebel, F. (1967). *Friedrich Froebel: A Selection from his writings*. Cambridge University Press.

Guattari, F. (1995). *Chaosmosis: An ethico-aesthetic paradigm*. Power Publications.

Guattari, F. (2008). *The three ecologies*. Continuum.

Hart, R. (1997). *Children's participation: The theory and practice of involving young citizens in community development and environmental care*. Earthscan.

Hickey-Moody, A. (2013). *Youth, arts, and education: Reassembling subjectivity through affect*. Routledge.

Kagawa, F., & Selby, D. (2010). Climate change education: A critical agenda for interesting times. In F. Kagawa, & D. Selby (Eds.), *Education and climate change: Living and learning in interesting times* (pp. 241–244). Routledge.

Kellett, M. (2005). *How to develop children as researchers*. Paul Chapman Publishing.

Kellett, M. (2010). *Rethinking children and research: Attitudes in contemporary society*. Continuum.

Lammes, S. (2019). The playful citizen: An introduction. In R. Glas, S. Lammes, M. de Lange, J. Raessens, & I. de Vries (Eds.), *The playful citizen. Civic engagement in a mediatized culture* (pp. 9–28). Amsterdam University Press. https://doi.org/10.5117/9789462984523/intro.

Lather, P., & St. Pierre, E. A. (2013). Post-qualitative Research. *International Journal of Qualitative Studies in Education, 26*(6), 629–633.

Lawler, J., & Patel, M. (2012). Exploring children's vulnerability to climate change and their role in advancing climate change adaptation in East Asia and the Pacific. *Environmental Development, 3*, 123–136. https://doi.org/10.1016/j.envdev.2012.04.001.

Lee, M. (2008). Curating the pedagogical scene. *Pedagogy, 8*(1), 194–198.

Lee, N. (2001). *Childhood and society: Growing up in an age of uncertainty*. Open University Press.

MacLure, M. (2013). The wonder of data. *Cultural Studies ↔ Critical Methodologies, 13*(4), 228–232.

MacRae, C. (2019). The red blanket: A dance of animacy. *Global Studies of Childhood*, https://doi.org/2043610619832899.

MacRae, C., & MacLure, M. (2021). Watching two-year-olds jump: Video method becomes 'haptic'. *Ethnography and Education, 16*(3), 263–278.

Manning, E. (2013). *Always more than one: Individuation's dance*. Duke University Press.
Manning, E. (2016). *The minor gesture*. Duke University Press.
Manning, E. (2019). Propositions for a radical pedagogy, or how to rethink value. In K. Leander, & C. Ehret (Eds.), *Affect in literacy learning and teaching: Pedagogies, politics and coming to know* (pp. 46–49). Routledge.
Manning, E., & Massumi, B. (2014). *Thought in the act passages in the ecology of experience*. University of Minnesota Press.
Massumi, B. (2002). *Parables for the virtual: Movement, affect, sensation*. Duke University Press.
Massumi, B. (2018). *99 theses on the revaluation of value: A post-capitalist manifesto*. University of Minnesota Press.
Morrow, V. (2008). Ethical dilemmas in research with children and young people about their social environments. *Children's Geographies, 6*(1), 49–61.
Murris, K. (2000). Can children do philosophy? *Journal of Philosophy of Education, 34*(2), 261–279.
Murris, K. (2016a). The Philosophy for Children curriculum: Resisting 'teacher proof' texts and the formation of the ideal philosopher child. *Studies in Philosophy and Education, 35*(1), 63–78.
Murris, K. (2016b). *The Posthuman child: Educational transformation through philosophy with picturebooks*. Routledge.
Punch, S. (2002). Research with children; the same or different from research with adults? *Childhood: A Global Journal of Child Research, 9*(3), 321–341.
Rolnik, S. (2017). The spheres of insurrection: Suggestions for combating the pimping of life. *E-flux Journal, 86*, 1–15.
Rosiek, J. L., Snyder, J., & Pratt, S. L. (2020). The new materialisms and Indigenous theories of non-human agency: Making the case for respectful anti-colonial engagement. *Qualitative Inquiry, 26*(3–4), 331–346.
Rousell, D. (2017). Mapping the Data Event: A posthumanist approach to art | education | research in a regional university. In L. Knight, & A. Lasczik Cutcher (Eds.), *Arts, research, education: connections and directions* (pp. 203–220). Springer International Publishing.
Rousell, D. (2019). Inhuman forms of life: On art as a problem for post-qualitative research. *International Journal of Qualitative Studies in Education, 32*(7), 887–908.
Rousell, D. (2020). A map you can walk into: Immersive cartography and the speculative potentials of data. *Qualitative Inquiry [special issue on "new directions in inquiry"]*. https://doi.org/10.1177/1077800420935927.
Rousell, D. (2021). *Immersive cartography and post-qualitative inquiry: A speculative adventure in research-creation*. Routledge.
Rousell, D., & Cutter-Mackenzie-Knowles, A. (2020a). A systematic review of climate change education: giving children and young people a 'voice' and a 'hand' in redressing climate change. *Children's Geographies, 18*(2), 191–208. https://doi.org/10.1080/14733285.2019.1614532.
Rousell, D., & Cutter-Mackenzie-Knowles, A. (2020b). Uncommon worlds: Towards an ecological aesthetics of childhood in the Anthropocene. In A. Cutter-Mackenzie-Knowles, K. Malone, & E. Barratt Hacking (Eds.), *Research handbook on childhoodnature: Assemblages of childhood and nature research*. Springer Nature.
Shaviro, S. (2009). *Without criteria: Kant, Whitehead, Deleuze, and aesthetics*. MIT Press.
Singh, J. (2017). *Unthinking mastery: Dehumanism and decolonial entanglements*. Duke University Press.
Skelton, T. (2008). Research with children and young people: exploring the tensions between ethics, competence and participation. *Children's Geographies, 6*(1), 21–36.

Sorin, R. & Galloway, G. (2006). Constructs of childhood: Constructs of self. *Children Australia, 31*(2), 12–21.

St. Pierre, E. A. (2011). Anything can happen and does. *Cultural Studies ↔ Critical Methodologies, 11*(4), 386–389. https://doi.org/10.1177/1532708611414670

St. Pierre, E. A. (2017). Deleuze and Guattari's language for new empirical inquiry. *Educational Philosophy and Theory, 49*(11), 1080–1089. https://doi.org/10.1080/00131857.2016.1151761

St. Pierre, E. A. (2019). Post qualitative inquiry in an ontology of immanence. *Qualitative Inquiry, 25*(1), 3–16.

Stern, D. N. (2010). *Forms of vitality: Exploring dynamic experience in psychology, the arts, psychotherapy, and development.* Oxford University Press.

Tanner, T. (2010). Shifting the narrative: Child-led responses to climate change and disasters in El Salvador and the Philippines. *Children and Society, 24,* 339–351. https://doi.org/10.1111/j.1099-0860.2010.00316.x.

Taylor, A., Pacinini-Ketchabaw, V., & Blaise, M. (2012). Children's relations to the more-than-human world. *Contemporary Issues in Early Childhood, 13*(2), 81–85.

UNICEF (1989). A simplified version of the United Nations Conventions of the Child. http://www.unicef.org.au/Upload/UNICEF/Media/Our%20work/childfriendlycrc.pdf?utm_source=multiple&utm_medium=short_url&utm_term=&utm_name=crc. Accessed 15 Jan 2018.

Vygotsky, L. S. (1986). *Thought and language.* MIT Press.

Vygotsky, L. S. (1997). *Educational psychology.* St Lucie Press.

Vygotsky, L. S. (2004). Imagination and creativity in childhood. *Journal of Russian and East European Psychology, 42*(1), 7–97.

Wake, S. (2008). 'In the best interests of the child': juggling the geography of children's gardens (between adult agendas and children's needs). *Children's Geographies, 6*(4), 423–435.

Whitehead, A. (1978). *Process and reality.* The Free Press.

Whitehead, A. N. (1967). *Adventures of ideas.* The Free Press.

Whyte, K. P. (2018). Indigenous science (fiction) for the Anthropocene: Ancestral dystopias and fantasies of climate change crises. *Environment and Planning E: Nature and Space, 1*(1–2), 224–242. https://doi.org/10.1177/2514848618777621.

Winnicott, D. W. (1988). *Babies and their mothers.* Free Association Books.

Winnicott, D. W. (1989). *Playing and Reality.* Routledge.

Wohlleben, P. (2016). *The hidden life of trees: What they feel, how they communicate – Discoveries from a secret world.* Black inc.

Wood, E., & Attfield, J. (2005). *Play, learning and the early childhood curriculum* (2nd ed.). Paul Chapman Publishing.

Young, T., & Cutter-Mackenzie-Knowles, A. (2020). Posthumanist learning: Nature as event. In A. Cutter-Mackenzie-Knowles, A. Lasczik, J. Wilks, M. Logan, A. Turner, & W. Boyd (Eds.), *Touchstones for deterritorialising socioecological learning: The anthropocene, posthumanism and common worlds as creative milieux* (pp. 27–48). Palgrave McMillian.

4

AN ECOLOGICAL AESTHETICS OF CHILDHOOD

In addressing the need for a more robust engagement with aesthetics in posthumanist studies of childhood and nature, this chapter makes some tentative steps towards an ecological aesthetics of childhood that is grounded in Whitehead's speculative philosophy. We make the case for a new aesthetics of childhood that is responsive to the accelerating social, technological, and environmental changes to planetary experience in the Anthropocene epoch. Our approach foregrounds the singularity of children's aesthetic experiences as relational-qualitative "intensities" that alter the fabric of nature as an ever-shifting field of vibrational experience. We draw on examples of children's environmental photography and poetry to elaborate an ecological aesthetics capable of attending to more-than-human sensibility, feeling, thought, and even fabulation. We argue that every event in a child's life is an uncommon and unrepeatable occasion through which the more-than-human world of nature is sensed, felt, perceived, and experienced differently.

Our turn to ecological aesthetics was driven by the emergence of children's creative practices in the Climate Change and Me project. As we witnessed children's encounters with climate change through film, photography, drawing, poetry, and fiction, we realised the need for a theoretical framework capable of grasping the aesthetic intensities of children's experience within these collectively generated events. Taking Whitehead's aesthetic order of nature as our starting point, we began to develop an ecological aesthetics orientated towards the "uncommon" dimensions of children's experience in and of the Anthropocene epoch (Rousell and Cutter-Mackenzie-Knowles, 2020). This theoretical work enabled us to recast aesthetic notions of image, sensibility, perception, and causality in relational terms, while also acknowledging the singularity of childhood experiences as creaturely becomings that produce new potentials for environmental thought and behaviour. Breaking with an aesthetics of distanced judgement according to transcendent criteria, this chapter proposes an ecological aesthetics of experience at no remove from life's

DOI: 10.4324/9781003336006-4

immediacy. This materialist, ecological aesthetics of childhood is attuned to the uncommon worlds through which children make aesthetic contact with particular creatures, environments, and modes of existence, and how the specific qualities and tonalities of this aesthetic contact are expressed through children's creative practices in response to climate change.

Materialist Aesthetics

As discussed in previous chapters, the mutual transformation of social, technological, and environmental systems has produced a posthuman condition in which figurations of childhood and youth are being drastically recalibrated under conditions of climatic, political, social, and ecological instability. This requires not only a reconsideration of children and young people's ecological relationships with nonhuman entities and environments but a retheorisation of children's aesthetic experience as coextensive with the "total environment" of nature as an open whole. This work engages posthumanism as inextricably connected with a fundamental rethinking of the aesthetic figure of the *human qua nature*, effectively dismantling lingering myths of the human individual as a bounded, rational, and independent agent. As Steven Shaviro argues, we need to learn to think of nature "without any residual anthropocentrism: that is to say, without exempting ourselves from it, and also without remaking it in or own image" (Shaviro, 2015, p. 216).

Samantha Frost, in her book *Biocultural Creatures*, further emphasises the need to rethink the very concept and nature of the "human" outside the inherited boundaries of human exceptionalism and anthropocentrism. "What we need in place of the fantasy of human exceptionalism is a different figure of the human, one that does not succumb to the conceits of old but also does not conceptually dissolve humans as identifiable agents and thereby absolve them of the crises that mark the Anthropocene" (Frost, 2016, p. 13). Frost shows how the porous, leaky images of human life offered by the postgenomic biosciences open up radical possibilities for critical inquiry in the posthuman humanities, arts, and social sciences. Emphasising the permeable composition of the human as *creature*, Frost argues for a "biocultural" reconceptualisation of human life as continuously assembled and re-assembled through the novel imbrication of social, technical, biological, and ecological processes. This chapter works to develop an ecological aesthetics of childhood that takes this radical permeability of human life into account, connecting Whitehead's (1978) speculative empiricism with Frost's (2016) new materialist account of biocultural life and bodily porosity. This turn to a radically *ecological* aesthetics is driven by the need to take account of children's experience as inseparable from the dynamic processes of the natural world, and thereby capable of sensing the processual dynamism of climate change in novel and transformative ways.

Affirmative engagements with aesthetics have often been lacking in posthumanist reconceptualisations of childhood studies and environmental education (Rousell and Cutter-Mackenzie-Knowles, 2020). This may be due to the eschewal of aesthetics that has accompanied the critical distancing of posthumanist studies of childhood from Enlightenment and Romantic aesthetic traditions, and their role in perpetuating purist, progressive, and sublime images of childhood and nature as idealised forms (Taylor, 2013, p. 13). We concur that such a break with transcendent aesthetics is both timely and necessary, as the humanising influences of Romantic idealism have remained default philosophical touchstones for contemporary environmentalisms (Morton, 2007), as well as "child in nature" movements associated with forest schools and related environmental education initiatives (Malone, 2016). In resisting the dominant tropes of humanistic ideals in environmental education and childhood studies, critical posthumanist studies have shown how representations of the "natural child" and "child in nature" have been used to perpetuate political regimes of racialisation, anthropocentrism, and human exceptionalism, effectively legitimising the very social and environmental injustices which they seek to counter (Malone, 2015; Taylor, 2013). The dominant cultural logic of a "distanced", Kantian aesthetics has also maintained ideal qualities (such as beauty) and capacities (such as art, language, and technology) as criteria for valuing certain qualities and capacities of the "human" above others, consequently positioning the external environment or "nature" as an objective correlate or backdrop for Western human progress, cognition, appreciation, and knowledge (Morton, 2013; Shaviro, 2014).

As Morton's (2007) deconstruction of the lingering dominance of Romanticism has shown, the dismantling of such idealist aesthetic regimes has been a core project of the arts for over two centuries. While Kantian aesthetics may continue to exert a stultifying influence on studies of childhood, education, and environment (Prout, 2005; Taylor, 2013), in the art world these idealisms have long since been replaced by the austerities of social realism, the machinic abstractions of modernism, the anti-humanist deconstructions of postmodernism, and the relational assemblages and spatio-temporal mobilities of contemporary art (Groys, 2008). Equivalently, contemporary work in art theory, cultural studies, and media studies has been largely influenced by Deleuzian, Whiteheadian, and new materialist aesthetics of immanence, sensation, relationality, affectivity, animality, and becoming (see, for instance, Hansen, 2015; Massumi, 2011; Shaviro, 2009). As Kennedy (2000, p. 30) notes, the materialist turn in aesthetics has decoupled beauty from ideals "consilient with goodness, the romantic, or transcendent notions", and reinscribed the beautiful within the "feeling of duration, movement, and continual process". We argue that posthumanist childhood studies can benefit from a more affirmative engagement with these ecological and new materialist turns in art, media, design, architecture, technology, and film studies over the last three decades (Rousell and Cutter-Mackenzie-Knowles, 2020). Donna Haraway's (2016) "new new synthesis" between the environmental arts, humanities, and sciences could provide a catalyst in this regard, drawing together studies of human and nonhuman aesthetics, ecologies, histories, affects, performances, technologies, and more (p. 63).

While we concur with Western scholars calling for an aesthetics that is radically environmental and inclusive of all forms of life and modes of existence (c.f. Haraway, 2016; Shaviro, 2009), we also note the particular resonance of such a (re)turn to aesthetics with Indigenous cosmologies and lifeways that have been in practice for millennia (Horton, 2017; Todd, 2015). Geo-ontological analyses of Indigenous art, philosophy, and culture by feminist scholars such as Grosz (2008) and Povinelli (2016) highlight an emerging sense of compatibility between the "traditional" aesthetics of Indigenous peoples and the "new" aesthetics proposed by today's speculative theorists, artists, and scientific practitioners (Rousell and Williams, 2020). We acknowledge that many Indigenous cultures have already been thinking and working through an ecological aesthetics for millennia, and offer a plethora of place-based and culturally responsive resources for grappling with the cosmopolitical challenges of social and ecological crises at the planetary scale (c.f. Bawaka Country et al., 2020; Dion and Salamanca, 2018).

Drawing on the geophilosophy of Deleuze and Guattari (1994), we might also thread a speculative history of ecological aesthetics through the Baroque territorialisations of the animal world, including the ritual performances and habitat constructions of mammals, birds, and myriad other forms of animal life (Rousell, 2019). And yet we acknowledge that the conditions for aesthetic experience in the contemporary world are *undeniably new*. As Hansen (2015) notes in his analysis of the experiential impacts of twenty-first-century media technologies, "we literally live in a new world, a world characterised by a vastly expanded and deterritorialised sensorium" (p. 161). There is no place on Earth that is unaffected by the inhuman proliferation of financial capital and the technical transmutation of life at planetary scale. The sheer number of human bodies continues to grow, even as the numbers of other Earthly creatures continues to decline. The cumulative sensing capacities of microcomputational media networks have become powerful agencies and elemental components of planetary existence (Gabrys, 2016). A global pandemic has spread from captive populations of nonhuman animals into human bodies, infecting mass human populations worldwide and crippling economies and health sectors. Nobody has ever experienced a world like we are collectively experiencing at present.

Aesthesis

There is an emerging sense that a more expansive, experimental, and promiscuous account of aesthetics is needed to grasp the experiential qualities of life under these radically altered conditions. Our turn to ecological aesthetics in this book can be situated within a veritable revival, what some have even called a *renaissance* (Gaskill and Nocek, 2014) of interest in aesthetics as the basis for speculative philosophy and new materialist theorisation (Debaise, 2016; Hansen, 2015). Often drawing extensively from philosophers such as Whitehead, Deleuze, and Guattari, scholars and artists associated with the speculative turn have brought aesthetics back into the centre of philosophical thinking and inquiry (Debaise, 2017; Kolozova, 2016; Shaviro, 2014). This speculative expansion of aesthetics to encompass both the cultural and the natural, the subjective and the objective, the artistic and the scientific is what we name an "ecological aesthetics". It is an aesthetics that operates across scales of space and time, from the microtemporal entanglement of quantum events to the geo-social movements of planetary epochs and evolutionary life processes (Yusoff, 2015).

Our particular (re)turn to aesthetics here is ecologically concerned with the circulation and exchange of felt relations between diversely situated bodies, communities, technologies, and environments. As Rousell and Williams (2020) write in this vein:

> An ecological aesthetics is concerned with the sensible qualities of relations between and amongst bodies, environments, societies, and technologies within complex assemblages that exceed the limits of human cognition and knowledge. The (re)turn to aesthetics is thus a (re)turn to the wonder of felt relation, to the sensibly distributed *nature* of experience as entangled with the wild variety of other creatures with whom we share our ecological worlds.
>
> *(Rousell and Williams, 2020, pp. 1605–1606)*

Rousell and Williams go on to note how this engagement with ecological aesthetics constitutes a return to the ancient Greek roots of the word "aesthetic":

> 1) *aisthetikos*, meaning "of or pertaining to αἰσθητά, things perceptible by the senses, things material, as well as perceptive, sharp in the senses"; and,
> 2) *aisthanesthai*, meaning "to feel, apprehend by the senses".
>
> *(Oxford English Dictionary, 2017, n.p.)*

In these two ancient definitions of aesthetics, there is a mutual inclusion between "things material" of the world and qualitative experiences of feeling and sensuous apprehension. The term *aesthesis* thus includes both the constructive act of sensation and the material constitution of that which is actually sensed perceived. Drawing on Whitehead's speculative empiricism in conversation with contemporary new materialism, Shaviro (2014, p. 156) argues that aesthesis constitutes the "primordial form of all experience, and thereby all action and relation". He describes the process of aesthesis as "contact at a distance", evoking a sensibility and sensitivity to worldly relation that occurs "in the realm of immanent, noncognitive contact" (p. 148). As a mode of aesthetic contact that collapses proximity and distance, aesthesis prefigures and preconditions the possibilities for knowledge and cognition, implying an immanent aesthetic connection with nature as the metaphysical ground for thought.

Consider, for example, a photograph of the sunset taken by Kairo, a 12-year-old researcher in the Climate Change and Me project (see Figure 4.1). By making contact with this image on the surface of the page we are also making contact with Kairo, and through the lens of her camera, with the event of a sunset taking place on a beach in Northern NSW Australia a number of years ago. Aesthesis is the movement of feeling that passes through each of these thresholds of experience regardless of the distance between events or the passage of time. Each touch or "grasp" carries

FIGURE 4.1 Photograph of the sunset taken by Kairo, age 12, CC+Me Researcher.

the singular feeling of an event through the image, what Whitehead (1978) terms a "vector of feeling", or "prehension". Our commitment to developing an ecological aesthetics of childhood invites this kind of intimate, creaturely, and sensorial aesthesis to shape our analysis and discussion of children and young people's research on climate change. We believe such an aesthetics can help foreground the *creatureliness* of children's felt relations and bodily capacities, while remaining attuned to incipient ethical and political ecologies as they emerge with/in environments that simultaneously surround, permeate, sustain, constrain, and animate children's everyday lives.

The Bifurcation of Nature

Taking up aesthetics in this ecological key requires a certain resistance to what Whitehead diagnosed nearly a century ago as "the bifurcation of nature". As a mathematician and logician, Whitehead (1964) was interested in formulating an alternative approach to the empirical sciences that resisted the bifurcation of nature into "the nature *apprehended* in awareness and the nature which is the *cause* of awareness" (pp. 30–31). He argued that the separation of an objective world of natural causality from a subjective world of aesthetic experience and rational interpretation constituted a futile and misguided "attempt to exhibit natural science as an investigation of the cause of the fact of knowledge" (p. 31). And yet the dualist bifurcation of nature, largely inherited from the Enlightenment era, has remained a universal "ground truth" for both logical positivism and humanistic claims to a transcendent or "free" imagination. In this sense, the bifurcation effectively posits nature as equally "there" to be discovered by the inquiring method of the scientist and interpreted by the imaginative sensibilities of the poet. Both cases assume that nature can be held at a distance and submitted to evaluation by an inquiring and fervent human mind. The bifurcation of mind from nature, values from facts, and meaning from matter provides equal support for the most reductive positivism and the most humanistic idealism, which conveniently lay claim to opposing sides of the bifurcation and are thus able to flourish independently of one another (Rousell, 2020).

Whitehead's process philosophy attempts to overcome this bifurcation by proposing a speculative empiricism which treats nature as composed of vibratory events, or processes of becoming, which are self-causing and ultimately aesthetic in composition. "Everything perceived is in nature", Whitehead writes in his early book *The Concept of Nature*. "We may not pick and choose. For us the red glow of the sunset should be as much a part of nature as are the molecules and electric waves by which men of science would explain the phenomenon" (1964, p. 20). To resist the bifurcation of nature is to affirm that the red glow of the sunset is both *what* it is objectively and *how* it appears subjectively, at one and the same time. The colour red, the sensation of red, the feeling of red, the intensity of red, the idea of red, the molecular materiality of red, past associations with red, the way that a certain shade of red appears at this particular time and place: these all become elements of nature as coextensive with aesthetic experience, or *aesthesis*.

If we take nature as an aesthetic patterning of vibratory events coming into expression, as Whitehead does, then every natural event is always an "event here", inhabiting its own relational perspective of the universe as a "concrete slab of nature" that generates its own vibratory space, time, and causality (p. 53). Each singular, emplaced perspective is a novel expression of aesthetic experience which also belongs to the whole of nature as a total event or situation. As Isabelle Stengers writes in her reading of Whitehead's concept of nature:

> The event that provides you with a point of view belongs to the great impersonal web of events. Your standpoint testifies to the whole of nature, is connected to the whole of nature, even if it takes on the particular meaning that is required by the interpretation of perception as yours.
>
> *(Stengers, 2011, p. 65)*

Thinking and working across the bifurcation requires a topological sense of part-whole relations that connects each singular event of experience with nature as an open totality. Without an aesthetic sense of a totality "brooding over its parts" (Whitehead, 1968), nature continues to *appear* bifurcated because the human mind interprets the external world as the final cause of its experience and the sensate object of its claim to knowledge. Whitehead's speculative empiricism offers a logical and consistent system for exploring what an unbifurcated aesthetic of nature instantiate through living practices, while also providing critical conceptual tools for sensing the cuts that habitual bifurcations make in practices of all kinds. Our argument here is not that the bifurcation of something called "mind" from something called "nature" is logically (or ethically) right or wrong, and that everyone should suddenly adopt an unbifurcated view of the world. Rather, the argument is that nature *always already includes* whatever bifurcations humans and other agents might potentially make of it. Our question then becomes: How does the relative bifurcation of *un*bifurcation of nature come to matter in children's creative practices as aesthetic encounters with climate change?

In the following sections, we address this question through an extended engagement with Whitehead's philosophy in conversation with collaborative works of art, media, creative writing, and theorisation produced by children and young people. Comments from the student researchers further contextualise these works as dynamic processes of ecological aesthesis that reveal valuable insights into the singular, affective impressions of climate change in children and young people's lives. In cultivating Whitehead's concepts of creativity, becoming, affective feeling, and prehension, our aim is to establish an ecological aesthetics of contemporary childhood which refuses the Western bifurcation of culture from nature, body from environment, and art from life. This approach pivots on a non-anthropocentric account of aesthetic affectivity, or what Whitehead simply terms "feeling", as the constitutive process through which bodies of all kinds (human and nonhuman) come into being, enter into relations, and sustain their relational existence through time. This approach differs from many contemporary affect theories that take human emotions and bodily capacities as the primary units of analysis (c.f. Ahmed, 2004), while also deviating in certain ways from Deleuze-inspired affect theories associated with vital materialisms (c.f. Bennett, 2010). Here our thinking is more closely indebted to recent developments of Whitehead's affect theory in the philosophy of art (Massumi, 2011; Manning, 2013) and media (Hansen, 2015; Shaviro, 2009), where affectivity is encountered at the germinal level of change-taking-place, prior to any categorical distinction between human or nonhuman elements or components of the event (Rousell, 2021). We elaborate this, below, by attending to the relations between *common* (public, extensive, ecological) and *uncommon* (private, intensive, aesthetic) worlds of creaturely experience, and how such relations between the common and the uncommon offer new ways of understanding children's experiences of climate change in the twenty-first century.

The Creatures That Become

As we described in Chapter 4, the collaborative production of works of art, media, fiction, and poetics throughout the Climate Change and Me project was led by children's speculative practices arising in the nexus between their everyday contexts and the existential threat of global climate change. The *Past Now Future* exhibition emerged as a novel platform to carry children and young people's climate change imaginaries and creative practices into the public domain. And yet we were left wondering about how to account for the rippling layers of affective turbulence that this co-creation and installation of works set into motion. How do we account for the ways that children's practices shift the affective *conditions* through which climate change is sensed, felt, thought, experienced, and articulated? How, for instance, does a child's creative encounter with an insect or a glacier *shift the conditions* through which climate change can be grasped and responded to?

The speculative scheme detailed by Whitehead (1978) in *Process and Reality* gives us a unique logical framework for responding to these questions, as it offers one of the most intricate and precise accounts of non-anthropocentric affect and relationality in the history of philosophy. The speculative scheme attributes aesthetic experience to all things that exist, and is constructed around what Whitehead terms "actual entities" as mutually implicated processes of becoming, or "occasions of experience". Every actual occasion of experience is proposed as a "creature of becoming" that is self-causing (*causa sui*), and sustains its own existence through affective encounters with other creaturely occasions of experience. Each creature of becoming is *aesthetic*, to the extent that it feels the relations of the actual world differently with each passing moment, expressing the principle of "novel togetherness" that Whitehead calls "creativity" (p. 21). Equally, all creatures in Whitehead's scheme are *ecological*, in the sense that their aesthetic experience is what produces the extensive physical environment that all organisms co-depend on to survive. "The actual occasions are the creatures which become", Whitehead (1978, p. 35) writes, "and they constitute a continuously extensive world".

For Whitehead, reality can be said to be generated by *creatures* because all actual entities *experience* the world through an aesthetic process of ontogenesis. Creatures are not the bounded organisms or things that we might recognise in the world, but the processual activity that generates those bodies through their relations with one another, with each passing moment. In other words, all things are the processual *concrescence* of creaturely occasions of experience, where experience is by no means relegated to or defined by *human* experience, but extends equally, but differently, to all things in their specificity (Rousell, 2021). The differences between discrete bodies are therefore considered differences of degrees, qualities, manners, and intensities of creaturely experience, rather than ontological differences in kind (Rousell and Cutter-Mackenzie-Knowles, 2020). Consider, for instance, the differences between a glacier and a child. The glacier advances and withdraws according to atmospheric variations in global temperature and precipitation, shifting states between solid, liquid, and gas. The child grows and sustains a trajectory of life through different modes and intensities of experience from the glacier, exhibiting different degrees, values, and capacities for aesthetic sensitivity to shared environmental conditions (see Figure 4.2).

Both the child and the glacier are changing constantly at different speeds, durations, and intensities, but these rates and degrees of change are linked and co-determining (change changing). These differences are differences in how bodies become and endure. They are differences in modes and manners of experiential activity,

FIGURE 4.2 Photograph of Franz Josef glacier in New Zealand by Sam, 12-year-old CC + Me researcher.

which suggests that the difference between the experience of a glacier and that of a child is *axiological*, rather than ontological (Robinson, 2009). This is because these differences in creaturely experience are distinguished by differences in *becoming* (as aesthetic modes, styles, or manners of existence) rather than differences in *being* (as the ontological essence or substance of existence). Different creatures generate their own *value* through the relational modes and manners of experience that they collectively instantiate. This value cannot be predetermined according to a hierarchy of being because it is pragmatically tied to different manners of concrete activity: the glacier withdraws, the child advances. This activity does not presuppose a hierarchy of value which assumes that the experience of the "child" or the "human" is more important or interesting than that of the glacier. Instead, patterns of axiological value must be generated anew every time through the relational unfolding of creaturely experience, with each instantiation of value contributing to novel contrasts and tonalities of experience as the shifting ground of mutual co-existence.

For Whitehead, this more-than-human production of value through aesthetic experience extends across all scales of existence, from the cosmological to the quantum, making his philosophy a system of "pan-experientialism" highly compatible with process thinkers such as Bergson and Deleuze. As Robinson (2009, p. 22) notes, each of these three thinkers foregrounds "the implicit aesthetic genesis, order and creative organisation of experience, with experience here understood as enlarged and not simply equivalent to human experience". We also note the compatibility of Whitehead's philosophy of organism with recent insights from the life sciences, which have highlighted the extent to which genetic, cellular, and bodily activities are dynamically modulated by both immediate and transgenerational changes in environmental conditions (Frost, 2016). As Protevi (2013) argues, contemporary advances in the biological sciences continue to confirm the process philosophies of Whitehead, Bergson, Simondon, and Deleuze by acknowledging the environmental porosity, plasticity, and permeability of sensing bodies and the environments they co-inhabit as the condition for life (see also Alaimo, 2010). Frost (2016) notes that the profound permeability of body and environment confirmed by the new life sciences also demands a rethinking of "culture" beyond anthropocentric terms of discourse and symbolic worldmaking. She suggests that the notion of "culture" might be understood as a verb rather than as a noun, referring to active processes of *cultivation* through which bodies and environments co-emerge, grow, develop, intersect, and eventually perish. This notion of cultivation as "culturing" would encompass a wide spectrum of material and discursive elements that enter into the composition of bodies and environments of every kind. As Frost explains:

> To think of culture in terms of cultivation enables us to incorporate in our thinking simultaneously and sometimes variously aesthetic, commodified, composted, cruciferous, arboreal, fermented, fractured, institutional, linguistic, metaphoric, normative, oceanic, organic, patterned, political, representational, subjective, trashed, treasured, violent ... worlds we inhabit.
>
> *(Frost, 2016, p. 4)*

Culturing, in other words, comes to describe the wildly diverse and divergent milieus through which more than human lifeworlds come into being and endure across time. We might now return to the examples of the glacier and the child in order to better understand what Whitehead means by the "creatures that become". The child steps onto the glacier with camera in hand. Both child and glacier are composed of *creatures*: they are accretions of all the relations they have experienced since they came into existence, including the auto-affective experiences of their own physical bodies, spaces, and times. In Whitehead's terms, they are distinct creaturely "societies", each with their own relational "routes of inheritance" and shifting horizons of futurity (Whitehead, 1978).

The child-society and the glacier-society are composed of very different creatures of experience, with divergent routes of inheritance from one moment to the next. Yet they inhabit, if only for a moment, a shared milieu or "culture" as they enter into an event of mutual encounter. This can be exemplified in Figure 4.3, through a photograph by 12-year-old researcher Sam during a visit to Franz Josef glacier in the South Island of New Zealand/Aeteroa. Sam describes how shifting patterns of holes, pools, and crevices would form in the glacier overnight, lasting only for a few days before dephasing into a new formation. In this event of encounter, as child puts camera to eye and begins to photograph the glacier, an event

FIGURE 4.3 Photograph of Franz Josef glacier by Sam, 12-year-old CC + Me researcher, accompanied by the caption: "Holes like this in the glacier would appear every so often, though they would not last long. It was predicted to melt and fall apart within a few days".

occurs that does not belong to the child or the glacier but to an *aesthetic patterning of child-glacier experience*. Child and glacier are sharing elements of each other's actual worlds, and this sharing of elements constitutes a novel aesthetic pattern made possible by a shared milieu of experience, or "culturing" in Frost's non-anthropocentric sense of the term.

There is, in this encounter of child and glacier, nothing static. Nothing is fixed. The pattern is moving, fluctuating, varying from itself through linked rates of change in processual activity. Each variation provokes a shift in the aesthetic texture of experience, altering the patterns of contrast to varying degrees of intensity. The singularity of this felt variation of relation is what Whitehead refers to as "prehension". Prehension is the way that creatures sense, perceive, and incorporate the experiences of other creatures through the aesthetic feeling and intensity of relation. Bodies become *what they are* by prehending or "feeling" other bodies, including their own prior states (Shaviro, 2009). This means that every encounter is irreducibly new. No child steps onto the same glacier twice. A *creature is what it feels*, and this feeling introduces novelty into the world. Every event is a new experience, a new affective relation, a new aesthetic contrast, a new culture.

Common and Uncommon Worlds

One of the key provocations that Whitehead's affect theory makes in relation to children's creative practices is the radical expansion of aesthetic experience to encompass the whole of nature (1926, p. 91). What we find within Whitehead's non-anthropocentric philosophy is a system of thought in which all orders of existence (including the biological, the social, the conceptual, the epistemic, the semiotic, the political, the ethical, the spiritual, the actual, and the virtual) are circumscribed within the aesthetic order of nature. As Judith Jones (1998, p. 20) notes, this conception of aesthetic order is coextensive with the weight and intricacy of causal complexity as it proliferates through ecological relationality and accretes extensively over time. In this respect, Whitehead's notion of aesthetic order offers a unique set of conceptual tools that connect the intensive worlds of individual bodies with the vast ecology of the universe as an open whole or "total environment".

Whitehead's (1967a, p. 53) aesthetic order of nature refers to a common universe of experience as differentially felt and transformed with each passing moment. This aesthetic order includes at least some trace of every event that has ever occurred, from the Big Bang to the flap of an insect wing in some remote corner of the Earth. And yet it also includes all the virtual potentialities and entangled sensibilities that particular creatures inherit from previous events, the evidence of which is empirically apparent in everything that surrounds us all the time. "The actual elements perceived by our senses are in themselves the elements of a common world", Whitehead (1967b, p. 88) writes. Everything that exists is always already immersed within this common world of "colours, sounds, and other sense-objects, related in space and time to enduring objects such as stones, trees, and human bodies" (p. 89). This sense of immersion in the aesthetic order of nature is implicit in the empirical act of sensation as well as the shared "sense" of an extensive world inhabited by all the creatures in the universe. As Whitehead notes, the "direct interrogation of our perceptive experience" consistently reveals that "we seem to be ourselves elements of this world in the same sense as are the other things which we perceive" (p. 89). The rustling of leaves outside the window, the low hum of the fridge, a child calling out from the room next store. All of these sensible elements of experiences are direct, empirical confirmation that this common universe exists, just like we do.

Thinking as a mathematical physicist and logician, Whitehead figures this common world of nature as a vibratory continuum that ripples across all scales and proportions of space-time, from the sub-atomic to the cosmological. This vibratory continuum is real, extensive, continuous, and objective to the extent that it includes every creature and occasion of experience that actually exists as a vibratory pulse in the field of nature. It constitutes a common world because it includes all physical elements and actualities of experience that belong to the concrete conditions of experience. And yet this multiplicity of elements and conditions of experience are also changing, with every moment, in ways that are not reducible to anything common. Even the merest movement of an electron introduces a new creaturely feeling into this continuum. These shifting conditions are what Whitehead calls the "actual worlds" of the creatures who become what we might call the biocultural habitats and or milieus of creatures and the societies they populate (Frost, 2016). These actual worlds are uncommon because they are specific to the perspective that each creature contributes to the universe. Whitehead suggests that each of these worlds is composed of *private* intensities of experience that are unique to the events through which they occur. While these uncommon worlds are not isolated from one another, they are not "common" because they are always being felt differently by the particular creatures and societies who actually experience them. For instance, in the encounter of child and glacier there is an extensive "region" of space-time that is shared, but there is also an intensity of aesthetic experience that is not. In the event of encounter, the child and the glacier each experience the event differently. Indeed, according to Whitehead, everything (from a child, to a stick insect, to a mushroom, to an asteroid) feels its actual world differently (uncommonly) as a particular perspective of the universe as a world in common. This grants every creature its own "private" intensity as a singular process of aesthetic individuation, inhabiting its own *uncommon* world in relation to a *common* world determined by how other creatures feel their interpenetrating worlds in different ways. It is these different ways of feeling and valuing events that, as we discussed in the previous section, introduce novelty into the world.

Working through this differential between common and uncommon worlds has helped us tease out some initial propositions for understanding children's relationships with climate change through their creative practices (Rousell and Cutter-Mackenzie-Knowles, 2020). Firstly, *children never experience the same actual world twice.* Each experience constitutes a novel synthesis of affective relations from a series of disjunctive elements (molecules, bodies, societies, places, times, materials, concepts, feelings, perceptions, discourses).

This aligns in some ways with Taylor's (2013) observation, drawn from work with Arrente Aboriginal peoples of Central Australia, that "children *are* their worlds. It is the particularities and distinctive features of these worlds that makes children who they are" (p. 115). However, we suggest that the very distinctiveness of children's actual worlds is indicative of their *uncommonality* as processes of individuation with their own private intensities of feeling and bodily capacities for aesthetic experience. This is not to say that children's worlds are disconnected from the common world of nature, but on the contrary, that each felt connection *with* the common world of nature is singular and new in its qualitative intensity. Every moment in the life of a child is an uncommon and unrepeatable instance through which the common world of nature is felt, perceived, and experienced differently. This became palpable in the Climate Change and Me project through the thousands of photographs that children submitted to the online archive. Figure 4.4, for example, shows a photograph by Grace (12-year-old researcher) exploring what she calls the "tiny worlds" of everyday interactions between insect and plant life in her back yard.

Second, according to Whitehead (1978, p. 35) the common world of nature is not continuous in itself but, rather, *becomes* continuous through the ongoing proliferation of uncommon worlds in all their teeming intensities and entanglements of creaturely relations. The continuity of the common world of nature relies on the discrete, uncommon intensities of experience to continue its extension. This means that a child's process of becoming is fundamentally creative at the level of natural causality: the child feels the actual world in a way that is *qualitatively* different from other creatures, and, in doing so, adds her experience to the extensive continuum of nature. In this way, the extensive continuum of nature both *produces* and *is produced*

FIGURE 4.4 Photograph by Grace, age 12, CC+Me Researcher, with the annotation: "The everyday lives of bugs and their food sources - the little worlds that exist around us".

by the intensive qualities of experience that constitute children (and everything else) through the "creatures that become". Everything a child experiences contributes to this sheer ongoingness of nature, while at the same time irrevocably changing the conditions for what nature can become. This means that there is nothing intrinsically normal or natural about childhood. The aesthetic order of nature offers no predetermined model for evolution or development, or transcendent criteria for evaluating what happens. The aesthetic order of nature is simply composed of the patterns of felt experience that all the creatures (organic and inorganic) have inherited from the past, and continue to feed differently into the future in order to endure. The experience of a child is just one of those patterns.

The Whir of Worldly Sensibility

In a common world in which everything both senses and is sensed through differential qualities of aesthetic experience, what comes to matter is the distribution and processing of aesthetic sensibility within more-than-human political ecologies and collectives. This necessarily involves a reconceptualisation of perception that participates creatively in the construction and patterning of aesthetic experience, rather than being a passive process that simply "feeds" data into the nervous system through an input-output model (Rousell, 2019). In addressing the problem of perception, Whitehead describes how "we find ourselves in a buzzing world, amid a democracy of fellow creatures" (1978, p. 50). This buzzing world of creaturely activity "expresses the relation of causal interaction between entities and their actual worlds" (Jones, 1998, p. 20).

As we described in the previous section, the causal interactions between creatures and milieus constitute the aesthetic order of nature as an extensive, vibratory continuum, or common world. The idea that the very consistency, order, and causal infrastructure of nature are altered and redistributed by the aesthetic experiences of events offers a basis for a new ecological aesthetics of childhood. It means that children literally reconfigure the virtual potentials of nature through intensive processes of individuation, creating new possibilities for life through the dynamic relations between a milieu of interiority and a milieu of exteriority (Simondon, 2017). Whitehead (1978) describes this process of individuation as the constructive functioning of "creatures", whereby the trafficking between intensive and extensive milieus, between private and public, uncommon and common, involves a creative synthesis of differential elements that produces the singular world of a living organism or society.

Media theorist Mark Hansen (2015) describes this ontogenetic process as "the distribution of worldly sensibility" (p. 3). Drawing on Whitehead's philosophy to reconceptualise the operations of twenty-first-century media, Hansen defines worldly sensibility as the causal infrastructure of the vibratory processes that operate primarily outside the thresholds of human sense perception and consciousness (p. 2). The cells of our own bodies, for instance, are constantly responding to the surrounding environment in ways that we are only vaguely aware of at the conscious level, if at all. These processes operate through what Whitehead (1978, 1929) calls the mode of "causal efficacy", which he contrasts with the perceptual mode of "presentational immediacy".

Where the mode of presentational immediacy involves the sensuous perception of vivid values through qualia such as colours, sounds, textures, tastes, and smells, causal efficacy is a function of the underlying causal infrastructure of nature as a vibratory continuum. Manning and Massumi (2014) argue that both presentational immediacy and causal efficacy are present to varying degrees in animal perception, making it a matter of *how* these modes of perception and sensibility achieve various fusions, syntheses, and vicarious interchanges. Perception comes down to practices of backgrounding and foregrounding that bring different patterns of felt experience into varying degrees of resolution and vagueness.

Presentational immediacy gives bodies access to the sensuous qualities of their social environments. It reveals clearly recognisable objects of perception, but it lacks temporal duration, thickness, and subtlety. In this sense, presentational immediacy is only the province of sufficiently complex animals who are capable of experiencing "the clear and distinct representation of sensations in the mind of a conscious, perceiving subject" (Shaviro, 2014, p. 23). Although Whitehead (1927, p. 23) describes such perception as "vivid, precise, and barren", it is through this representational mode of perception that "the world discloses itself to be a community of actual things, which are actual in the same sense as we are" (p. 21).

Causal efficacy, on the other hand, refers to the "vague (nonrepresentational) way that entities affect and are affected by one another through a process of vector transmission" (Shaviro, 2014, p. 23). It is thus more accurately described as a mode of aesthetic *sensibility*, or aesthesis, rather than a mode of perception (Hansen, 2015). Causal efficacy opens the range and distribution of sensibility to the entire aesthetic order of nature, and places all things that exist on the same ontological footing. Causal efficacy suggests that even our most clear and distinct perceptions are mediated through complex causal processes that lie outside consciousness, including those processes associated with our own bodily capacities.

> We see the picture, and we see it with our eyes; we touch the wood, and we touch it with our hands; we smell the rose, and we smell it with our nose; we hear the bell, and we hear it with our ears; we taste the sugar, and we taste it with our palate.
>
> *(Whitehead, 1927, p. 50)*

In these examples, causal efficacy is associated with what Whitehead (1978, p. 81) calls "the withness of the body", a position which implicates the body and its capacities within the elemental distribution of *worldly sensibility* across the whole of nature. When immersed in the mode of causal efficacy, "the inflow into ourselves of feelings from enveloping nature overwhelm us … the presentations of sense fade away, and we are left with the vague feeling of influences from vague things around us" (Whitehead, 1978, p. 176).

As Hansen (2015) explains, the vague, haunting sensations of causal efficacy are indexed to "the *material causal linkages* that exist and have force outside of the realm of impressions, ideas, and ideas of impressions – outside the domain of sense perception proper" (p. 49, emphasis in original). In bringing the vibratory continuum of nature into contact with the perceptual impressions and affective intensities of individual creatures and societies, Whitehead's theory of prehension effectively fuses aesthetic experience with the material causal infrastructure of nature's spatiotemporal dynamisms (Protevi, 2013). This radically extends the act of perception into a visceral spectrum of sensibility that, at the limit, is plugged into the entire aesthetic order of nature as an open totality of self-causing creatures and self-organising causal, material infrastructures (Figure 4.5).

Meyer (2005) notes that most cognitive and neuroscientific studies of perception have yet to even consider the possibility of sense perception in the mode of causal efficacy. By focusing on neuronal processes taking place in particular regions of the brain, neuroscientific accounts of experience largely ignore the vague, non-specific influences of material causalities that are not directly identifiable as a discrete stimulus.[1] Consider, for example, the image of the ladybug with a water droplet in a photograph that Grace (12-year-old researcher) theorises as a "tiny world" (Figure 5). Attention to presentational immediacy would only account for how the image is processed into recognisable "chunks" which enable the brain to "recognise" the form of an insect, sitting on a grass stem, with a drop of water on its back. No current neuroscientific models would be able to account for the vague, seething expanse of blurry green in the background, which bears witness to the play of

FIGURE 4.5 Photograph by Grace, age 12, CC+Me Researcher.

worldly sensibility across the vibratory continuum of nature as a rippling, open whole. In other words, neuroscience fails to account for how every "tiny world" also includes the entire universe as part of its ontogenesis. As we explore further in Chapter 5, taking worldly sensibility into account requires us to consider how any image, perception, or experience is connected to the whole of nature as a creative process. As you doze in a tiny world amidst the incessant hum of cicadas on a summer afternoon, you might also hear the vague hum of worldly sensibility in the background. A whole buzzing universe of natural causal processes seething just below the thresholds of conscious sense-perception.

Sensing Climate Change

By taking a deep dive into Whitehead's speculative empiricism as a theoretical touchstone for this book, our aim is to assemble alternative understandings of children's aesthetic experiences of climate change in the twenty-first century. Our speculative wager is that Whitehead's focus on ecological aesthetics opens a different range of affective, perceptual, and artistic registers for sensing the planetary impact of climate change on children at the level of worldly sensibility. Hansen (2015) describes this type of approach as "radically environmental", to the extent that bodies and environments are considered coextensive with the aesthetic fabric of the vibratory order of nature. Yet he also stresses the extraordinary impact of new media technologies as technical agents that are increasingly altering the weave of this experiential fabric, and therefore warping the extensive continuum of space-time itself (de Freitas, 2018; de Freitas et al., 2019).

There is a similar argument at play in Hansen's claim to a radically expanded sensibility through twenty-first-century media technologies, and our project's attention to children's creative activity as modalities of sensing climate change at the creaturely level. Certainly digital media technologies played a significant role in our project, with iPads, digital cameras, and online interfaces enabling children and young people to sense, capture, exchange, reconfigure, and proliferate vectors of aesthetic feeling across multiple spaces and times. Yet our sustained investment in the concept of the *climate child imaginary* draws out a different angle on Hansen's notion of worldly sensibility, one that is grounded in analogue environmental sensation, and not necessarily reducible or accountable to the digital as its primary contemporary expression. This became particularly evident through the archive of poems that children generated over the course of the project. Consider the lines of this poem written by nine-year-old researcher Marie:

> As the new snow in the old world
> We can drench, once the snowflakes melt

Marie's image of new snow falling onto the old world renders, with simple and elegant precision, the aesthetic process of becoming that Whitehead elaborates in *Process and Reality*. Each experience feels the old world differently, adding itself to the settled world like fresh, falling snow. This new world always relies on the old in the creative process of producing the new. In the second line of the poem, Marie describes the "drenching" of experience as the new snow melts, again evoking Whitehead's sense of the "perpetual perishing" of each drop of felt experience as it passes into the next occasion. Considered together, Marie's lines gesture towards the possibility of extinction due to climate change, evoking the terrifying possibility of "drenching" once the world's ice has completely melted. We see these lines beautifully crystallising a child's aesthetic encounter with climate change by connecting her intensive world of childhood experience with the worldly sensibility of nature as an open totality. And yet, as Marie's poem suggests, the all-embracing relations of nature's worldly sensibility can also feel truly suffocating when they can no longer sustain our current forms of life.

We highlight the extraordinary insight of Marie's poetic lines in order to foreground children's creaturely, aesthetic capacities to directly sense planetary changes without reliance on the digital or the computational. Marie told us how she worked for weeks over school holidays trying to get the words to "feel just right" for this poem. She talked about wrestling day and night with the vague, haunting sense of climate change seeping into her everyday life experiences, passing through the walls of her bedroom and infecting her thoughts and feelings. From the perspective of the ecological aesthetics that we unfold in this chapter, we can see how this poem emerges through a process of *aesthesis* that enables Marie to make aesthetic contact with an open totality of nature regardless of scale or distance. In Whitehead's (1967b, p. 83) terms, her poem "grasps the whole of nature in the tonality of the particular instance".

Marie's poem expresses a cosmological sense of nature's dynamism, while touching on the felt immediacy of aesthetic intensity in the passing present. It exemplifies a non-cognitive, aesthetic sensitivity to climate change that is simply not (yet) possible through digital or computational means, which are almost exclusively built on binary logics and cognitive processing models. This is not to position a child's creaturely sensibility *against* that of a machine but, rather, to suggest that a machine capable of *organic* worldly sensibility and thought has yet to be invented. Even so-called machine learning algorithms simply execute functions based on a binary cognitive logic of accumulation, and thus rely on a drastically delimited image of what constitutes learning (Mackenzie, 2017), and indeed, "what constitutes legitimate thought processes" (Goodman, 2020, p. 49). We therefore argue that sensing climate change cannot just (or even primarily) be a question of techno-scientific machinery, algorithms, and data. Nor is it necessarily a question of decentring or dissolving human experience. Indeed, such a move becomes moot with the acknowledgement that there are millions of centres of experience comprising the human body alone (Whitehead, 1978), many of which would not, in any normative sense, be considered "human" (Frost, 2016). Rather, we suggest that sensing climate change is fundamentally a question of how creatures make aesthetic contact with a radically changing world in all its teeming wildness and complexity, and generate new patterns of felt experience that shape the contours of the future differently.

Our sustained attention to the ecological aesthetics of children's encounters with climate change seeks to open spaces for cultivating this kind of creaturely sensitivity and poetics, where *aesthetic experience* becomes the worldly instrument for sensing and responding to precariously shifting environmental milieus.

Note

1 In this respect, Whitehead's extension of perception into the mode of causal efficacy constitutes one of the most challenging contributions of his philosophical project to contemporary studies of mind and consciousness. Almost a century since the publication of Process and Reality, Whitehead's theory of perception offers a radical challenge to prevailing neuroscientific concepts of mentality and evolution, and the potential to read and interpret contemporary neuroscientific data differently (see Meyer, 2005).

References

Ahmed, S. (2004). *The cultural politics of emotion*. Edinburgh University Press.

Alaimo, S. (2010). *Bodily natures: Science, environment, and the material self*. Indiana University Press.

Bawaka Country, including Wright, S., Suchet-Pearson, S., Lloyd, K., Burarrwanga, L., Ganambarr, R., Ganambarr-Stubbs, M., ... & Maymuru, D. (2020). Gathering of the clouds: Attending to Indigenous understandings of time and climate through songspirals. *Geoforum, 108*, 295–304.

Bennett, J. (2010). *Vibrant matter: A political ecology of things*. Duke University Press.

de Freitas, E. (2018). The biosocial subject: Sensor technologies and worldly sensibility. *Discourse: Studies in the Cultural Politics of Education, 39*(2), 292–308.

de Freitas, E., Rousell, D., & Jäger, N. (2019). Relational architectures and wearable space: Smart schools and the politics of ubiquitous sensation. *Research in Education, 107*(1), 10–32.

Debaise, D. (2016). *Speculative empiricism: Revisiting Whitehead* (T. Weber, Trans.). Edinburgh University Press.

Debaise, D. (2017). *Nature as event: The lure of the possible*. Duke University Press.

Deleuze, G., & Guattari, F. (1994). *What is philosophy?* Columbia University Press.

Dion, S. D., & Salamanca, A. (2018). Enunciation: Urban Indigenous being, digital storytelling and Indigenous film aesthetics. *The Canadian Journal of Native Studies, 38*(1), 183–207.

Frost, S. (2016). *Biocultural creatures: Toward a new theory of the human*. Duke University Press.

Gabrys, J. (2016). *Program earth: Environmental sensing technology and the making of a computational planet*. University of Minnesota Press.

Gaskill, N., & Nocek, A. J. (2014). *The Lure of Whitehead*. University of Minnesota Press.

Goodman, A. (2020). The Secret Life of Algorithms: speculation on queered futures of neurodiverse analgorithmic feeling and consciousness. *Transformations, 34*, 50–70.

Grosz, E. (2008). *Chaos, territory, art: Deleuze and the framing of the earth*. Columbia University Press.

Groys, B. (2008). The topology of contemporary art. In T. Smith, O. Enwezor, & N. Condee (Eds.), *Antinomies of art and culture: Modernity, postmodernity and contemporaneity* (pp. 71–82). Duke University Press.

Hansen, M. B. (2015). *Feed-forward: On the future of twenty-first-century media*. University of Chicago Press.

Haraway, D. (2016). *Staying with the trouble: Making kin in the Chthulucene*. Duke University Press.

Horton, J. L. (2017). Indigenous artists against the anthropocene. *Art Journal, 76*, 48–69. https://doi.org/10.1080/00043249.2017.1367192.

Jones, J. A. (1998). *Intensity: An essay in Whiteheadian ontology*. Vanderbilt University Press.

Kennedy, B. M. (2000). *Deleuze and cinema: The aesthetics of sensation*. Edinburgh University Press.

Kolozova, K. (2016). Preface: After the "Speculative Turn". In K. Kolozova, & E. A. Joy (Eds.), *After the "Speculative Turn": Realism, philosophy, and feminism* (pp. 9–17). Punctum Press.

Mackenzie, A. (2017). *Machine learners: Archaeology of a data practice*. MIT Press.

Malone, K. (2015). Posthumanist approaches to theorising children's human-nature relations. In K. Nairn, P. Kraftl, & T. Skelton (Eds.), *Space, place and environment, geographies of children and young people* (pp. 185–206). Springer.

Malone, K. (2016). Reconsidering children's encounters with nature and place using posthumanism. *Australian Journal of Environmental Education*, 32(1), 42–56.

Manning, E. (2013). *Always more than one: Individuation's dance*. Duke University Press.

Manning, E., & Massumi, B. (2014). *Thought in the act: Passages in the ecology of experience*. University of Minnesota Press.

Massumi, B. (2011). *Semblance and event: Activist philosophy and the occurrent arts*. MIT Press.

Meyer, S. (2005). Introduction: Whitehead now. *Configurations, 13*(1). https://doi.org/10.1353/con.2007.0010.

Morton, T. (2007). *Ecology without nature: Rethinking environmental aesthetics*. Harvard University Press.

Morton, T. (2013). *Hyperobjects: Philosophy and ecology after the end of the world*. University of Minnesota Press.

Oxford English Dictionary (2017). Aesthetic. https://en.oxforddictionaries.com/definition/aesthetic. Accessed 20 May 2017.

Povinelli, E. A. (2016). *Geontologies: A requiem to late liberalism*. Duke University Press.

Protevi, J. (2013). *Life, war, earth: Deleuze and the sciences*. University of Minnesota Press.

Prout, A. (2005). *The future of childhood: Towards the interdisciplinary study of children*. Routledge.

Robinson, K. (2009). *Introduction: Deleuze, Whitehead, Bergson—Rhizomatic connections*. Springer.

Rousell, D. (2019). Inhuman forms of life: on art as a problem for post-qualitative research. *International Journal of Qualitative Studies in Education, 32*(7), 887–908.

Rousell, D. (2020). Cosmopolitical encounters in environmental education: Becoming-ecological in the intertidal zones of Bundjalung National Park. *The Journal of Environmental Education, 52*(2), 133–148.

Rousell, D. (2021). *Immersive cartography and post-qualitative inquiry: A speculative adventure into research-creation*. Routledge.

Rousell, D., & Cutter-Mackenzie-Knowles, A. (2020). Uncommon Worlds: Towards an ecological aesthetics of childhood in the Anthropocene. In A. Cutter-Mackenzie-Knowles, K. Malone, & E. Barratt Hacking (Eds.), *Research handbook on childhoodnature: Assemblages of childhood and nature research*. Springer Nature.

Rousell, D., & Williams, D. (2020). Ecological aesthetics: New spaces, directions, and potentials. In A. Cutter-Mackenzie-Knowles, K. Malone, & E. Barratt Hacking (Eds.), *Research handbook on childhoodnature: Assemblages of childhood and nature research*. Springer Nature.

Shaviro, S. (2009). *Without criteria: Kant, Whitehead, Deleuze, and aesthetics*. MIT press.

Shaviro, S. (2014). *The universe of things: On speculative realism*. University of Minnesota Press.

Shaviro, S. (2015). *Discognition*. Repeater Books.
Simondon, G. (2017). *On the mode of existence of technical objects* (C. Malaspina, & J. Rogove, Trans.). Univoval Press.
Stengers, I. (2011). *Thinking with Whitehead: A free and wild creation of concepts* (M. Chase, Trans.). Harvard University Press.
Taylor, A. (2013). *Reconfiguring the natures of childhood*. Routledge.
Todd, Z. (2015). Indigenizing the Anthropocene. In H. Davis, & E. Turpin (Eds.), *Art in the Anthropocene: Encounters among aesthetics, politics, environments and epistemologies* (pp. 241–254). Open Humanities Press.
Whitehead, A. N. (1926). *Religion in the making*. Cambridge University Press.
Whitehead, A. N. (1927). *Symbolism, its meaning and effect* (Barbour Page Lectures, University of Virginia).
Whitehead, A. N. (1929). *The function of reason*. Free Press.
Whitehead, A. N. (1964). *The concept of nature*. Cambridge University Press.
Whitehead, A. N. (1967a). *Adventures of ideas*. The Free Press.
Whitehead, A. N. (1967b). *Science and the modern world*. The Free Press.
Whitehead, A. N. (1968). *Modes of thought*. The Free Press.
Whitehead, A. N. (1978). *Process and reality*. The Free Press.
Yusoff, K. (2015). Geologic subjects: Nonhuman origins, geomorphic aesthetics and the art of becoming in human. *Cultural Geographies*, *22*(3), 383–407.

5

SURFACES OF EXPERIENCE

In Chapter 4, we introduced Whitehead's speculative empiricism as the basis for assembling an ecological aesthetics capable of accounting for children and young people's shifting affective experiences of climate change, and how these experiences in turn shift the conditions for how climate change can thought, felt, and imagined. Whitehead's philosophy is particularly suited to this task due to its focus on the primacy of relation and affectivity as the basis for acts of living, thinking, perceiving, and imagining. Here we engage further with Whitehead's theory of feeling in order to establish a framework for grasping the creative evolution of children's image-making practices in the Climate Change and Me project. This is important in order to resist the conventions of representational and cognitive reductions of images to inert schematic models of information transmission and signification. Reading Whitehead's speculative empiricism through a posthumanist philosophy of the image, Ivakhiv notes that

> how we image the world, and how we take up those images in our acts of imaging, is both how we add to the world and how we constitute ourselves in the process. There is no safe space into which we can retreat without being defiled and transformed by our acts of imaging.
>
> *(Ivakhiv, 2018, p. 155)*

By leaning into Whitehead's theory of feeling, this chapter aims to theorise children's photoparagraphs as vectors or "carriers" of environmental experience as wells as attractors or "lures" for cultivating alternative climate change imaginaries. This allows us to open the image to a movement beyond its representational content and form, where we encounter an enfolded intricacy of feelings in the material depths of imagistic surfaces (Bruno, 2014; Marks, 2018).

DOI: 10.4324/9781003336006-5

In Whitehead's philosophical vocabulary "feeling" is synonymous with what he terms "prehension". As discussed in Chapter 4, prehension characterises how bodies take account of their actual worlds as a manifold of potentials for feeling. Whitehead (1978) describes two primary types of feeling which are always co-implicated within any occasion of experience. *Physical feelings* are the feelings of extensive causality, which is essentially how bodies affect and are affected by entering into mutual complexes of co-determining vibratory relations. "Thus a simple physical feeling is one feeling which feels another feeling" (Whitehead, 1978, p. 236). *Conceptual feelings* involve a sense of valuation and orientation towards potential events and relations, as well as what Whitehead terms "eternal objects", such as colours and numbers that make aesthetic "ingressions" into feeling. Conceptual feelings are characterised by what Whitehead calls "appetition", as both an "urge toward the future" and a "principle of unrest" which creatively conditions and mixes with the physical feelings associated with self-preservation and endurance. For example, Whitehead describes thirst as "an immediate physical feeling integrated with the conceptual prehension of its quenching … thirst is an appetite toward a difference- toward something relevant, something largely identical, but something with genuine novelty" (p. 33). Such entanglements of conceptual and physical feelings produce a wild variety of "hybrid" prehensions through which mental and physical relations become indissociable. Images are prime examples of hybrid prehensions, as patterned compositions of contrasting colour, shade, and tone that provide "vectors" for channelling physical and conceptual feeling into the future.

Vectors of Feeling

Importantly for our discussion here, prehension (both physical and conceptual) always takes on a "vector character" in Whitehead's thought. This is not simply a linear connection between discrete points, such as between a perceiving subject and a perceived object, or between points in the landscape or on a map. Rather, each prehension constitutes a vector of feeling that originates in some specific, determinate feeling from the past and terminates in the constitution of a future feeling. Through the passage of each vector of feeling there is a transference of physical energy (vibratory waves of matter in the extensive continuum) as well as a transference of what Whitehead terms the "touch of mentality", which "clothes" the physical feeling with a sense of aesthetic value and qualitative tonality.

Mentality, in Whitehead's terminology, is simply another word for conceptual feeling. And yet as Manning (2013) explains, the term "mentality" is often misleading due to the prevalence of cognitivist or phenomenological perspectives which assume a unitary, bounded human consciousness as the sole model of mental activity.

> Mentality is perhaps the wrong word for this intensive process, this "organ of novelty", or "urge beyond", for despite this not being the case, mentality … still sounds as though it is in the mind or in consciousness … We might

> therefore simply call it "thinking-feeling," emphasising how it is an activity in the event that co-composes with the occasion's physicality to create, in the act, a contributory more-than that emphasises how novelty is a process of thought in the doing.... The force of appetition ... could be another good term for mentality in that it emphasises the hunger of a process which opens the occasion to novel motifs, activating in the occasion a "factor of anarchy".
>
> *(Manning, 2013, p. 216)*

By situating processes of thinking-feeling in *events* rather than in minds, Whitehead offers a theory of affectivity which aims to account for how *nature thinks*. This has foundational implications for grasping the constructive functioning of space and time, not as primordial categories or containers for thought, but as conditionings of thought in the act. If thinking-feeling is essentially relational and always in transition, then space and time are also conditional to how an experience is physically and conceptually felt. This is to reject what Whitehead (1967b, p. 58) calls the "fallacy of simple location", which insists that physical "bits of matter" and mental "subjects" can be located absolutely in fixed points of space and time. Rather, Whitehead suggests that space and time themselves can be considered "basic forms of affectivity; they cannot be pre-assumed, but need to be constructed in and through the process of experience" (Shaviro, 2009, p. 60). Space-time becomes relative to vectors of feeling, in the sense that each process of feeling inherits the past and urges itself into the future, transitioning "from a beyond which is determinate and pointing to a beyond which is yet to be determined" (Whitehead, 1978, p. 230). The intervening "touch of mentality" is what orientates and aims the vector of feeling towards its terminus, adding a "factor of anarchy" that resolves into a novel feeling never before felt. Thus, vectors of feeling continuously introduce novelty into the world.

Feeling is also what connects space and time back to the body, where the body is posed "not as a stable category but as a creative vector of experiential space-time" (Manning, 2012, p. 12). The body becomes an instrument for the mutual sensitivity of conceptual and physical feelings that channel the passage of space-time through a vector of experience. This also means that classical categories of spatio-temporal distance and proximity are collapsed. That which exists "over there" beyond the body can be felt "right here", inside the bodily complex of affective relations and tonalities, as the constructive functioning of a process of feeling that is indeterminate, and yet conditional to the world's becoming (Rousell, 2021). Vectors of feeling not only connect bodies across multiple spatial distances, they also connect memories of past occasions across multiple times.

These vectors of memory flow in different temporal directions (past to present, present to past, future to past) and are intermixed constantly in the nexus of affective modulations and transductions that constitutes a bodily event. For an example, consider Grace's photograph of two glistening mushrooms emerging from a background of lush green moss (Figure 5.1). Grace took this photograph while exploring the rainforest remnants near her grandparents' house in the Northern

FIGURE 5.1 Photograph of mushrooms by Grace, 11-year-old researcher in the CC+Me project.

NSW hinterland. In the act of looking at Grace's image, you feel it *here* in the event wherever you find yourself reading this book. The image activates a vector of feeling that extends from the child's past encounter with mushrooms into a future feeling that is only now being realised. The bodying of this feeling may urge you to think back to previous occasions where you encountered mushrooms, or walked with a child in a lush forest valley. It may also activate an appetite to walk outside later this week, looking for mushrooms or other creatures. The power of the image, in this example, is revealed as an intricate nexus of feelings that catalyses and animates events across many different scales and temporalities.

The Environ/mental Imaginary

What would be involved in the act of sensing an image, then, if we continue with Whitehead's theory of feeling? Going back to Whitehead's early critique of the bifurcation of nature, we witness an attempt to unify the appearance of objects with how they actually operate at the level of vibrational causal processes. Consider Whitehead's example discussed in Chapter 4, which asked us to feel the red glow of the sunset while simultaneously acknowledging the play of microscopic particles, the retina, and brain activity that cause the red glow of the sunset to appear in perception. Whitehead argues that the appearance and the cause of perception are inseparable aspects of the passage of nature, refuting the "sensationalist doctrine" that posits the mind as internal interpreter of external stimuli. From Whitehead's

FIGURE 5.2 Photograph of glacier melt in New Zealand/Aotearoa by 12-year-old researcher Sam.

unbifurcated view of nature, the appearance of the sunset is *in* the cause, and the cause of the sunset is *in* the appearance. We can apply this unbifurcated view of nature to consider the appearance, for example, of Sam's photograph (Figure 5.2) of a melting glacier as inseparable from the causal infrastructures that enable the image to appear.

There are, of course, many material causal processes at work in the rendering of the appearance of any image *in* sensation. Depending on the depth and granularity of our analysis, we can follow the chemical, climatological, and atmospheric conditions that produce the singularity of the melting glacier as a passing event, occurring from its own standpoint, within a particular place and time. We can also attempt to follow the tendencies and appetitions that lead Sam to witness this glacial event and select a particular framing and moment to snap a photograph while peering through the viewfinder of a digital camera. The camera transduces voltages in response to the photons arriving at its sensor pixels within a specified window of time. The sensor pixels are converted into chroma (colour) and luma (brightness) data and compressed into a digital codec, such as JPEG. As a compressed packet of code, the JPEG image can then be transferred from the camera's storage to any digital device and online media sharing platform, such as the Climate Change and

Me "Hangout" space. When the encoded digital file is opened, the image appears on the surface of any device, and can also be transferred into print via artefactual media such as this book.

While this brief material account emphasises an image's physical itinerary from Sam's singular encounter with a glacier to the pages of this book, Whitehead's philosophy suggests that there is something more at play in the proliferation of images than mechanistic, physical processes of transfer. As we discussed above, each prehensive phasing of an image is not only physical, but also *conceptual*, in the sense that there is a factor of mentality at play in each movement of an image along a vector of feeling and valuation. This intensive dimension of the image could be understood in terms of an "environ/mentality", where the mentality or "sense" of an image is located in the event of its movement through an environmental field. This environ/mentality of the image is never complete and always changing, atmospherically, depending on the local conditions of prehension in particular places and times.

Adrien Ivakhiv (2018) offers an insightful account of this imagistic process in his book *Shadowing the Anthropocene*, which applies Whitehead's philosophy to reconceptualise the agency of images within ecological cultures and imaginaries. Ivakhiv describes images as "immanent, active, moving, and engaging" within a processual reality that is, quite literally, *made* of aesthetic images.

> Reality is imagistic... It is prehension all the way down, with crucial differences arising in the qualities of each prehension—how they reshape what they (selectively) inherit, how they move with that which they have gathered.... Image is, in this sense, everything. But image is never pure, true, accurate, or complete. It is never a perfect representation of something else, because it is always slipping from what it is imaging. It is always in motion, between what is imaged from and where it images to. Image, in this definition, is a creative reconfiguration of elements perceived or prehended, a configuration that, once satisfied, is passed on to the world for further refiguration.
>
> *(Ivakhiv, 2018, p. 153)*

Ivakhiv's account of an "imagistic reality" emphasises the dynamic play of configuration and refiguration that animates each vector of feeling as a particular image of the universe. For Ivakhiv, it is this environ/mental play of images participating in each other's realities that constitutes the ongoing production of an affective "imaginary". This perspective is important in our broader argument, throughout this book, for the significance of constructing alternative climate child imaginaries with children and young people. Images can be considered, in this respect, the concrete elements of sense and sensation of which imaginaries are composed. And imaginaries, for their part, direct the affective currency of images within a teeming meshwork of vectors of feeling that simultaneously inherit the past and create the future. In the case of the climate emergency, our reading of images needs to move beyond the representational in order to recognise how images constitute the affective fabrics of social worlds. Perhaps this means shifting the emphasis and function of images away from "raising awareness" of climate change, towards rerouting the affective currency of images that constitute ourselves and our environments.

Maybe what comes to matter, in a world of melting glaciers and mass extinctions, isn't the visual content of the image but the *environ/mentality* it feeds into the future. In theorising the image as a vector of environ/mentality that animates children's encounters with climate change, the image is untethered from the visual in order to embrace the wider swathe of sensibility associated with causal efficacy, as the worldly sensibility of nature itself. No longer tethered to visual perception, the image becomes a haptic interface, a folded surface of experience, a movement of feeling, an event of change perceived from its own standpoint within the whole of nature.

Proliferating Images

Drawing on the philosophical discussion above, we are interested in the ways that climate child imaginaries are, quite literally, *built* from images that prefigure and reconfigure affective constellations of lived events. In our work with child researchers, speculative practices of image-making exerted a powerful gravitational pull. From the initial phases of the project, photography emerged as a creative research method that opened up spaces for children and young people to experiment with noticing, sensing, witnessing, framing, interpreting, and composing alternative climate change imaginaries. This process was set in motion through the project's emergent research design and child-framed methodology, which evolved by working collaboratively with children and young people within a "research playspace", as discussed in Chapter 3.

We were amazed to find that many of the CC+Me researchers took up photography as a primary method of inquiry, as we witnessed an inundation of over 4,000 images uploaded to the project website in the first year of the project alone. We found that children were actively exploring their local environments, discovering images of felled trees, nests of baby birds, messy bedroom computer desks, backyard chickens, roadside trash, and the everyday lives of insects and frogs. Yet we also found that children were taking their photographic research with them on journeys during school holidays. We began to receive images, for example, of a receding glacier landscape in New Zealand, squirrels in a park near the Eifel tower, and fish underwater near the Great Barrier Reef. The children's vigorous engagement with photography was framed by the shared goal, as described in Chapter 3, of creating an immersive public exhibition called *Past Now Future* that ended up reaching nearly 10,000 members of the general public. As the photographs began to sample an increasing range of the world's environments, we also began to assemble an extraordinarily diverse and detailed image of how children were sensing and perceiving the imbrication of climate change in all types of places and times. We began to think about each image as, quite literally, an affective sampling of the Earth's environ/mental transmutation as captured from the singular perspective of a child.

As thousands of images piled up in our digital archive, we were increasingly faced with the problem of how to sense (and make sense) of these images analytically. As we progressed through the analytic stages of the project, we hosted a series of creative ThinkTank workshops where the children worked through the data en masse. In order to deal with the masses of photographic data, we spread thousands of images out on the floor of a large gymnasium and simply invited the children to rifle through them and select the ones that "called out", "spoke", "provoked", "mattered", or "glowed" (MacLure, 2013). This resulted in a process of collectively curating the photographs down to a series of 200 images that the children felt expressed their experiences of climate change, and of life in the Anthropocene more broadly. During this process, the children were also invited to write messages on the backs of the photographs. The messages were given a combined function of expressing the felt qualities of children's perceptions of the images, while also resonating those feelings for a public audience. In some cases children formed groups and began to create a series of photographic "essays" that were eventually published in the *Past Now Future* exhibition catalogue and, more recently, in the ChildhoodNature Companion (Widdop Quinton et al., 2020).

Surfacing Children's Image-Making Practices

In this section we discuss several examples of how children took up the invitation to use photography as a research method in the Climate Change and Me project. Rather than interpreting the images as representations of what children think or know about climate change, we follow the intensive production of images as immersive surfaces that implicate children within their shifting environments. Building on our philosophical work in Chapter 4, we understand the surface of the image to be a composition of creaturely prehensions, or what Whitehead calls "vectors of feeling", that opens onto the vibratory continuum of worldly sensibility. Our intention is to plumb the intensive depths of the image's surface, moving away from linguistic and optical frames of reference to embrace a tactile materiality that fuses the immediacy of aesthetic feeling with the vague penumbra of worldly sensibility that surrounds each event. Through our focus on images as sites of aesthetic contact and felt relationality, our research essentially comes to involve "the production of surfaces, their multiplication and their consolidation" (Deleuze, 1990, p. 125). In this way, we see the surface of the image as a nexus between the uncommon worlds of children as singular creatures and the vibratory continuum of nature as a common world of material causality, relationality, sensibility, and extension.

Our discussion begins with a series of digital photographs taken by children, which we understand as prehensions of children's actual worlds as they exist(ed) in the places and times of the photographs. The first photograph was taken by an 11-year-old researcher named Grace, as she walked alone in the wet sclerophyll forest near her grandmother's house in a remote area of Northern NSW (Figure 5.3). Moistened by a heavy rain that had fallen over previous weeks, the forest was teeming with fresh explosions of fungi, vegetation, and insect life as Grace collected

FIGURE 5.3 Photograph taken by Grace (age 11) during a morning walk in wet sclerophyll forest.

a series of highly detailed and saturated images. The second photograph was taken by 12-year-old Kairo, who made her way to a remote coastal location in Northern NSW just before dawn (Figure 5.4). These images capture the Pacific Ocean's horizon in the liminal moments between night and day, such that the surfaces of water and sky become washed with colour values too subtle and blended to name.

These images both emit a certain dark brilliance, a gradient of contrasts that results in a particular intensity of aesthetic feeling emanating from the surface of the photograph. In this first analytic pass over the images, we are aware of values of colour, light, shadow, and texture resolving into recognisable forms on the surfaces of the photographs. The presentational immediacy of atmospheric qualities within a distributed environmental field resolve into clearly delineated spatial relationships: tiny mushrooms growing on a branch, an ocean reflecting the sky.

As Manning and Massumi (2014) note, there is a certain aesthetic enjoyment or "entertainment" that accompanies the presentational immediacy of perception prior to the resolution of the field's surface effects into recognisable forms and spatial relations. Entertaining the immediacy of the environment also brings our attention to the contemporaneous nature of events that occur relationally, and yet independently, within the field of experience. There are interpenetrating values, such as colour and moisture, which are shared between the ocean and the sky in Figure 5.4, a resonance across surfaces. This is to entertain the surface effects of the relational field itself as an environmental dance of aesthetic elements, a mode of perception perhaps closer to the experiences of autistic people or of hallucinogenic states rather than neurotypical consciousness (Manning, 2013).

FIGURE 5.4 Photograph by Kairo (age 12) just before dawn at Angel Beach, NSW.

But even before the surface of the image resolves into recognisable forms and affordances (mushrooms you could pick, an ocean you could hear, smell, and dive into), the sense of an image has already been infused with a vague mode of sensibility associated with causal efficacy. Attending to images in the mode of causal efficacy reveals that there is much more occurring in these photographs, both perceptually and causally, than what can be seen within the frame. In the first image we can barely sense Grace's damp foray into the rainforest, and share in her increasing wonder as she photographs the wild variety of fungal constellations appearing at her feet. In the second image we might catch a vague, penumbral sense of Kairo knee-deep in the waters of Angel Beach at dawn, trying to steady the camera as she darkly contemplates rising sea levels and human extinction. In each of these images *we sense what the child senses*, and what *the lens of her camera senses*, within a particular duration of time and space. Attending to causal efficacy gives us a haunting sense of what lies beyond the perceptible objects within the frame of the image, indexing the metaphysical weight of multiple spaces and times that condition the image directly, aesthetically, affectively, materially, causally.

The sensing body of the camera is crucial in this mediated configuration of child, image, and environment, as it captures and renders the sensory data of the occasion in ways that no human can do. As Hansen (2015, p. 24) argues, such media technologies are able to "capture a wealth of data from worldly sensibility- including data about our own implication in it- and feed this data forward into a future or just-to-come moment of conscious perception". In this sense, digital cameras give us access to the aesthetic order of nature in ways that conscious, human perception does not.

Again, we can understand the extrahuman sensing capacity of the camera as a function of what Whitehead (1978) calls the "vector-character" of prehension. The camera renders the image as a conduit or channel for vectors of feeling that extend from one unique perspective of the universe to another: sunlight falling on the forest floor, the roiling ocean at dawn, the wind-swept eyes of a child, the ink-saturated pages of a book. All of these and more can be considered events within an ecology of sensation that proliferates through an intricate meshwork of aesthetic prehensions. This sensory ecology of feeling makes it possible for you to encounter these images and words as you flip the pages of this book. There is not only a physical extension of imagery and words but also an intensive movement of feeling that passes through, and re-arranges, the affective currencies of the immediate environment. In this way, each image expresses a capacity to "channel the force of worldly sensibility" as the "pure potentiality" of the vibratory continuum of nature as the total environment through which every image passes (Hansen, 2015, p. 238). In sensing the image, we also grasp a vague sense of the whole of nature seething in the intensive depths of its surface.

Through this speculative analysis of the image as a conduit for vectors of feeling and worldly sensibility, we find that it is no longer the human/nonhuman distinction that really matters, or maybe even that this distinction is no longer possible. In the children's photography above, the images themselves show nothing human. And yet the body of the child is *felt* as a living society, invisible and yet implicated in the surface of the image. The sensible appearance of the image and the material conditions of its appearance are inseparable, such that the surface enables sensing bodies to make aesthetic contact with one another *at a distance*, through *aesthesis* (Shaviro, 2014, p. 148). This fusion of presentational immediacy with causal efficacy opens the image beyond any representational or discursive function. What we are dealing with instead are the differential relations between the forces of aesthetic appearance and material causality, between the resolution of perceptual elements within a sensory ecology and the material conditions through which that whole sensory ecology operates (as nature). Child-environment prehensions contract the elemental qualities of worldly sensibility into an image, channelling the field of potential experience to produce an intensity of contrast that is felt on the surface. This is what we mean when we say that the surface of the photograph has depth: it has depth of contrast, depth of feeling, depth of intensity, depth of sensibility, depth of movement, depth of beauty, and depth of potential.

Rematerialising the Image as Surface

What begins to emerge in this account of children's image-making is the sense of images as vectors of feeling that pass through, and reconfigure, the relations between bodies and environments. In our analysis workshops, as children rifled through thousands of images to find the ones that "speak", they are participating in the proliferation of vectors of feeling and the affective imaginaries that they engender in the social domain. Through this selection and annotation process, the child's sensation and interpretation of the image becomes *part of the image*, part of its affective channelling of feeling towards new encounters and imaginaries. This affective process of sensing images and *feeding these sensations forward* into the future is what, in conventional research terms, might be called "data analysis". However, in undertaking this collective analysis of images with children, we were not interested in representational themes or codes based on the frequency of particular images, words, or ideas. Rather, we were interested in how each image could be seen to enjoy a life of its own, as a particular perspective of the universe that is continuously introducing something new into the world. In this sense, our speculative approach aligns with recent post-qualitative movements in education and social research which abandon the representational concepts and conventions of traditional qualitative and arts-based inquiry (Rousell, 2019; St. Pierre, 2019).

By undertaking this speculative approach to data analysis, we also found that the massive proliferation of images was generating a complex event of felt *togetherness* through the co-production of a shared imaginary. In addition to the singularity of each image as a vector of feeling, the sheer *quantity* of images also presented a powerful affective force akin to what Massumi (2018) describes as "intensive magnitude". We wanted to develop a collaborative process that would enable public audiences to sense the collective force of this emerging climate child imaginary, in terms of both its qualitative intensity and its materiality as a tangible, physical process. Following the ThinkTank workshops, we realised that we need to develop a creative process that could effectively transduce the images from digital files into sensuous and tactile artefacts that people could actually hold in their hands. We explored a range of different technical processes, and eventually settled on the use of encaustic beeswax as a medium that could effectively "transfer" digital images into tactile analogue prints.

We developed this process of encaustic image transfer through extensive trial-and-error experimentation, working closely with children to design and assemble a collaborative, image-based artwork for the *Past Now Future* touring exhibition. This process effectively archived the images in an encaustic wax medium that would preserve them for millennia, with examples of Egyptian, Greek, Roman, and early Byzantine encaustic artworks preserving detailed images from as early as 100 AD. Our choice of encaustic wax medium was influenced by this extraordinary material capacity for a naturally occurring substance, such as beeswax, to preserve images for thousands of years. Yet this material capacity also speaks to the intensive, symbiogenic bond between human image-making practices and the life-generating practices of nonhuman animals and plants. Bees are crucial pollinators of the world's plant populations as well as "sensors" of environmental change at planetary scale. Bees also exhibit a unique capacity for collective intelligence and sensitive, affective responses to changes in ecological conditions. By working with beeswax as a medium for transducing digital photographs into archival analogue images, children were essentially "collaborating" with bees in the ecological proliferation of images and the creation of new climate child imaginaries.

Our use of beeswax also introduced very particular aesthetic qualities and processes into the rematerialisation of the children's images. As shown in Figure 5.5, the wax transfer process dynamically alters the aesthetic texture, colour value, and coherence of the image by introducing a range of organic disruptions and tonal shifts. These changes in the aesthetic coherence of the image are intricately tied to the material process that we developed to achieve the wax transfer. The first step in the preparation process involved mixing pure beeswax with damar resin, which is derived from tree sap collected in the tropical rainforests of India and Southeast Asia. The resin enables the beeswax to harden without cracking once cooled, and is often used in techniques such as batik and other forms of wax printing. Once we had created the encaustic medium by dissolving the resin and wax, we poured the molten mixture onto a series of 200 plywood rectangles, aiming for approximately 5mm of wax coverage. A heat gun was then used to create a smooth surface of wax medium on the surface of each plywood tile. Two hundred digital images were then printed out in high resolution using a laser printer, one for each tile.

Once this extensive preparation was complete, we worked with the children to undertake the wax transfer process for each of the images. The first step of this process is to gently heat the wax surface of the tile using a heat gun. The surface should be warm but not hot, just under molten temperature. The digital image is then pressed face down on the wax surface of the tile. After applying extensive pressure to the image with the back of a spoon, the paper from the original print can be removed by repeated rubbing with fingers (see Figure 5.5). Small amounts

FIGURE 5.5 A photograph by Grace (age 11) with the annotation "the little worlds that exist around us" transferred using encaustic medium on plywood.

of water can also be used to help remove the bits of paper. The surface of the tablet is then carefully reheated to a molten state in order to lift the underlying wax medium to the rippling surface, fixing the image within a translucent layer of wax and resin. This is perhaps the most dynamic step of the entire process, as the image itself become molten and floats like water on the surface of the wax. Once the wax cools, what appears on the surface of the tablet is a ghostly monoprint of the digital image. The printing ink from the digital image is now suspended within an archival medium that can, under certain conditions, preserve the image for thousands of years (see Figure 5.6).

The wax transfer process demonstrates how the image can function not as a representation but as a surface that channels vectors of worldly sensibility through processes of material transduction and affective transformation. Through this ecological process, the image slides from one medium and milieu into another. The transfer process effectively *resingularises* a digital image through its deformation and reformation as an analogue image (Guattari, 2008). In other words, the wax transfer is a process of transduction from one medium to another, from a medium of digital code to a medium of analogue materials and bodily sensations. While digital images can be ceaselessly reproduced without changing in coherence, the wax transfer process creates a version of the image that is singular, unique, comparable only to itself: *uncommon*. "Think of each image receding into its deformation, as into a vanishing point of its own twisted versioning", Massumi (2002, p. 133) writes. Paradoxically, we see the image become more singular by deforming and differentiating, resingularising back on itself as "matter in analogue mode" (p. 135). Strictly

FIGURE 5.6 Rubbing the paper off the surface of the wax transfer once the ink has been embedded.

speaking, however, what our wax transfer achieves is a novel *fusion* of digital and analogue modes of imaging. In this respect, we might suggest that the digital code is still "suspended" in the wax medium, to the extent that the laser printing inks still carry a trace of the binary code as an imprint. Our imaging process thus gestures towards a speculative fusion of digital and analogue modes of existence, where the digital is artfully *suspended* in the analogue, rather than determining and controlling the conditions of possibility for analogue life (see Figure 5.7).

This transductive fusion of digital and analogue imaging practices is important for our broader project, both aesthetically and by extension, politically. As discussed in Chapter 4, we are invested in exploring the relational intensities of children's climate change imaginaries as distinctive and singular processes, rather than mass aggregations. Moreover, we see the need to revalue analogue modes of thought and aesthetic sensibility due to the extraordinary reliance of digital systems on binary code and cognitive reductions of thinking and learning to computational mechanics. One of the ways that our process revalues the analogue is by valuing the organic deformations and contortions of image achieved by the intensive temperature fluctuations of the wax transfer. By heating the surface of the image to melting point, the precarious virtuality of the digital image is brought to the fore. The digital image can no longer be considered pure, neutral, coherent, complete, or static when its rematerialisation in analogue form is scarred by the process of deformation. Sensible qualities of the image are lost in the process of transfer, but other "mutant sensibilities" are gained (Kennedy, 2000). By approaching images as vectors of feeling, rather than static or complete objects, our transfer process emphasises the

FIGURE 5.7 An example of wax transfer image creating a digital-analogue fusion of body and environment.

liveliness of what an image includes in its frame, but also what it leaves out, what it loses, and what it leaves behind. "A feeling bears on itself the scars of its birth", Whitehead (1978, pp. 226–227) writes. "It retains the impress of what it might have been, but is not".

Through the fusion of digital and analogue imaging achieved by the transfer process, we also notice that the image seems to implicate the body more thoroughly into the depths of its surface. The inherent fusion of bodies and environments becomes more palpable, more intricately embedded and folded together through a renewed tactility of the body–environment nexus. Perhaps the digital code of the photograph was only a temporary holding pattern between the body's environmental experience and the rematerialisation of the analogue image on the wax tablet. As a milieu for making aesthetic contact with children and their biocultural environments, the transfer process replaces the representational ocularity of vision with a tactility that crosses temporal and spatial divides. When we pick up a wax tablet with an image suspended in its surface, we can literally touch the child's sense of moving through the forest as we brush out fingers over the surface. This speaks to what Hansen (2006, pp. 70–71) has termed an "infratactility", or primordial tactility, that operates at the very "heart of the sensible". The wax transfer process intensifies the tactility of the image, enabling us to touch the image as a vector of feeling. While there is a loss in clarity and resolution of the image, there is a gain in analogue sensation and worldly sensibility. The image becomes more permeable, taking on the biocultural tactility of skin, flesh, and membrane, embodying the "sometimes fluid and sometimes disjunctive processes of responsive development, dissipation, intensification, and transformation" (Frost, 2016, p. 149).

In the *Past Now Futures* exhibition, which travelled to eight libraries and galleries across NSW in 2015, the 200 wax tablets created by children were composed into an interactive artwork titled *Pieces of the Earth* (see Figure 5.8). This title reflected our collective sense that each tablet offered an imaging of children's aesthetic experience of climate change as a singular "piece" or fragment of the Earth's environment in a state of dynamic transformation. Each tile was fitted with magnetic strips, and featured a message to the public written by children on the back. In the installation, we used a movable magnetic board to display the 200 tiles as a fluid montage of images that public audiences could actively touch, feel, sense, and re-arrange as they interacted with the work. This openness to relational interaction continued to preserve the liveliness and collective singularity of the images, refusing to superimpose any particular code, pattern, theme, or interpretation. Each encounter with *Pieces of the Earth* thus shifted the affective currents and trajectories of the work, with each touch reconfiguring the vectors of feeling and the co-production of new climate child imaginaries.

Cultivating the Art of Life

In this chapter we have taken some steps towards a reconceptualisation of images as vectors of feeling for weaving new climate child imaginaries. Drawing on Whitehead's

FIGURE 5.8 Children engaging with the collaborative artwork *Pieces of the Earth* as a fluid montage of surfaces.

speculative philosophy in conjunction with insights from contemporary media studies, we proposed a series of theoretical engagements with images as vectors that open onto the environ/mentality of nature as an open totality. This involved a further series of distinctions between presentational immediacy and causal efficacy as modes of perception and worldly sensibility, with an emphasis on the fusion of children's perceptual experiences with the causal infrastructure of natural systems and dynamic spatiotemporal processes. We then extended this ecological-aesthetic framework through an analysis of children's environmental photographs as part of the Climate Change and Me project, including the analysis of digital photographs in relation to the analogue rematerialisation of images through a wax transfer process. Using aesthetic concepts of feeling, contrast, sensibility, and surface, our analysis focused on developing the image as a site where digital and analogue fusions of sensibility can produce new, hybrid forms of social practice and imagining.

This brings us to the perennial question of how images come to *work* as works of art that alter the affective fabrics and currencies of the social. Does a child's photograph become art when it is curated and exhibited as such? Does the transductive process of the wax transfer qualify the image as art through its material and affective transformation? While we do not offer a definitive answer to these questions here, we suggest that the transfer process intensifies both the elemental and aesthetic qualities of the image, and in doing so, pushes the image to the threshold where image becomes art, and art becomes the transmutation of social life. To the extent that the image becomes more singular through an artful process, and that this process introduces qualitative changes in material flows and affective currencies, then we would say that it becomes art. Hence, it is not so much the image itself that we advance as art but the transformative processes that are both attributed to and unleashed by the image through forces of material production and the intensities of affective encounters. It is in this sense that we associate art with *artfulness*, as an aesthetic mode or manner through which life is lived and experienced within a more-than-human ecology of practices (Manning, 2016).

We also recognise artfulness as a way in which the aesthetic becomes *pedagogical*, inasmuch as a work of art is able to redistribute worldly sensibility at the levels of the social, the ecological, and the planetary. We therefore conclude this chapter with a pedagogical perspective, as we position artfulness as the cultivation of aesthetic sensibilities for practicing what Whitehead (1929) calls "the art of life". We see the art of life as a pedagogy for cultivating environmental awareness through the aesthetic selection, valuing, and recomposition of sensory elements in relation to bodily movement and perception. "For example", Whitehead (1929, p. 200) writes, "the mere disposing of the human body and the eyesight so as to get a good view of the sunset is a simple form of artistic selection. The habit of art is the habit of enjoying vivid values". Whitehead positions the body as the pivot for the art of life, an aesthetic practice of enjoying the sensible intensities, values, and contrasts that emerge at the nexus of interior and exterior milieus, uncommon and common worlds. "Each individual embodies an adventure of existence. The art of life is the guidance of this adventure" (Whitehead, 1967a, p. 39).

The art of life is an art of the creature, an art of the organism, an art of collectively attuning to an ecology of sensation through aesthetic techniques of existence. These are techniques that meld aesthetics with environ/mental materiality, such that the beautiful is no longer associated with judgement or taste, but with "a felt 'immediacy', 'force', or 'intensity' in process" (Kennedy, 2000, p. 31). Life becomes a fluid montage of images. Grace walking in the forest finding mushrooms at her feet, or Kairo wading into the waters of Angel Beach to take photographs at dawn: these are images in which bodies are guided by the art of life, such that body and environment become fused in the production of life itself as a work of art (Rousell and Fell, 2018). These images are occasions of vivid experience never to be repeated; they are the uncommon worlds of climate child imaginaries, each a unique perspective of the universe as felt differently, and yet together, with each passing moment.

References

Bruno, G. (2014). *Surface: Matters of aesthetics, materiality, and media*. University of Chicago Press.

Deleuze, G. (1990). *The logic of sense* (trans: M. Lester). Columbia University Press.

Frost, S. (2016). *Biocultural creatures: Toward a new theory of the human*. Duke University Press.

Guattari, F. (2008). *The three ecologies*. Continuum.

Hansen, M. B. (2006). *Bodies in code: Interfaces with digital media*. Routledge.

Hansen, M. B. (2015). *Feed-forward: On the future of twenty-first-century media*. University of Chicago Press.
Ivakhiv, A. (2018). *Shadowing the anthropocene: Eco-realism for turbulent times*. Punctum Books.
Kennedy, B. M. (2000). *Deleuze and cinema the aesthetics of sensation*. Edinburgh University Press.
MacLure, M. (2013). The wonder of data. *Cultural Studies ↔ Critical Methodologies, 13*(4), 228–232.
Manning, E. (2012). *Relationscapes: Movement, art, philosophy*. MIT Press.
Manning, E. (2013). *Always more than one: Individuations dance*. Duke University Press.
Manning, E. (2016). *The minor gesture*. Duke University Press.
Manning, E., & Massumi, B. (2014). *Thought in the Act: Passages in the ecology of experience*. University of Minnesota Press.
Marks, L. U. (2018). 'We will exchange your likeness and recreate you in what you will not know': Transcultural process philosophy and the moving image. In H. Vaughan, & T. Connely (Eds.), *The Anthem handbook of screen theory* (pp. 119–143). Anthem Press.
Massumi, B. (2002). *Parables for the virtual: Movement, affect, sensation*. Duke University Press.
Massumi, B. (2018). *99 Theses on the revaluation of value: A postcapitalist manifesto*. University of Minnesota Press.
Rousell, D. (2019). Inhuman forms of life: On art as a problem for post-qualitative research. *International Journal of Qualitative Studies in Education, 32*(7), 887–908.
Rousell, D. (2021). *Immersive cartography and post-qualitative inquiry: A speculative adventure into research-creation*. Routledge.
Rousell, D., & Fell, F. (2018). Becoming a work of art: Collaboration, materiality and posthumanism in visual arts education. *International Journal of Education through Art, 14*(1), 91–110.
Shaviro, S. (2009). *Without criteria: Kant, Whitehead, Deleuze, and aesthetics*. MIT Press.
Shaviro, S. (2014). *The universe of things: On speculative realism*. University of Minnesota Press.
St. Pierre, E. A. (2019). Post qualitative inquiry in an ontology of immanence. *Qualitative Inquiry, 25*(1), 3–16.
Whitehead, A. N. (1929). *The function of reason*. Free Press.
Whitehead, A. N. (1967a). *Adventures of ideas*. The Free Press.
Whitehead, A. N. (1967b). *Science and the modern world*. The Free Press.
Whitehead, A. N. (1978). *Process and reality*. The Free Press.
Widdop Quinton, H., Piersol, L., Rousell, D., Russell, J., Dezan, R., Shannon, T., et al. (2020). Becoming companions: Compositions of childhoodnature relation, sense, poetics, and imagining by children and young people. In A. Cutter-Mackenzie-Knowles, K. Malone, & E. Barratt Hacking (Eds.), *Research handbook on childhoodnature: Assemblages of childhood and nature research* (pp. 687–756). Springer, Cham.

6

CHILDREN OF AN EARTH TO COME

Introduction

Speculative fiction emerged over the course of the Climate Change and Me project as a salient practice for exploring children's climate change imaginaries and affective orientations towards the future. This speculative trajectory of inquiry was not something that we, as facilitators of the project, necessarily planned or conjured into being. As described in Chapter 3, our approach was to collectively gather a research playspace that enabled speculative inquiry to emerge spontaneously through the creative vision and skills of our young collaborators. In this chapter, we describe how speculative fiction was initialised and pedagogically cultivated as a creative research method by Jasmyne, a 12-year-old researcher who created an elaborate series of fictional characters, narratives, illustrations, and environments during her participation in the CC+Me project. We describe how Jasmyne's visionary practices and ideas opened up spaces for other children to create their own characters and adventures within an open fictional world, eventually evolving into a collaborative story-building experiment that has had wide-ranging impact through the CC+Me curriculum and digital network. We focus on Jasmyne's practices of character-building and world-building as techniques for mobilising an aberrant subjectivity through anarchic political encounters with the speculative and the unknown.

In theorising this viral proliferation of speculative practices and ideas, we connect Jasmyne's characterisations and narratives with Deleuze and Guattari's (1987, 1994) geophilosophical concepts of deterritorialisation, fabulation, and the minor politics of "a people and an Earth to come". This enables us to consider the irruption of speculative fiction in our research project as a critical posthumanist response to the geo-traumas of climate change through the fabulation of alternative futures.

DOI: 10.4324/9781003336006-6

The Endings of Worlds

Increasingly, speculative fiction is being taken up as a creative technique for navigating the philosophical, educational, artistic, and political implications of climate change in response to critical posthumanist and post-anthropocentric concerns (see, for instance, de Freitas and Truman, 2020; Ghosh, 2016; Gough, 2015; Haraway, 2016; Morton, 2013; Shaviro, 2015; Weaver, 2010). In their discussion of the impacts of climate change on the Earth System as a whole, Danowski and Viveiros de Castro (2017, np) argue that the near future has become increasingly impossible to predict, "if not indeed unimaginable outside the framework of science fiction and messianic eschatologies". Amitav Ghosh's (2016) marking of the Anthropocene as the "Great Derangement" also gestures towards the inability of existing modes of cultural production to grapple with the disorientating weirdness of climate change. Ghosh argues that climate change troubles the stylistic conventions of literary fiction, film, and performance genres that have relied on the climate as a relatively stable background against which human drama can play out. Literary or dramatic productions that even mention climate change, as Ghosh observes, are almost immediately relegated to speculative fiction, "as though in the literary imagination climate change were somehow akin to extra-terrestrials or interplanetary travel" (p. 7).

There is now an emerging body of literature on the use of speculative fiction to pursue a dynamic range of resonant trajectories of inquiry in educational research (c.f. de Freitas and Truman, 2020; Doyle, 2020; Ehret and Čiklovan, 2020; Jensen, 2020; Truman, 2019). The permutations of this turn to speculative fabulation are wide-ranging, including speculative feminist interventions in the English classroom (Truman, 2019), explorations of more-than-human response-ability through decolonial speculative fiction (Nxumalo, 2020), and the use of speculative design to co-develop digital platforms with young people (Ehret and Čiklovan, 2020; Rousell et al., 2021). We are now seeing speculative fiction being engaged not only as an ontological tool (Rousell et al., 2017), but also as a new empirical modality of inquiry into education futures (de Freitas and Truman, 2020), capable of probing, activating, and mobilising new trajectories of thought and inquiry amidst the planetary-scale transmutations of a rapidly evolving posthuman condition (Braidotti and Bignall, 2019). This involves grappling with the existential threat of the climate emergency, but also, as we've discussed in previous chapters, with the accompanying technical and biopolitical transmutation of planetary life and sociality.

Our project's turn to the speculative reflects not only an engagement with speculative philosophy and fiction as ecologically agentic and socially transformative modes of critical inquiry, but also an acknowledgement of Indigenous cultures that have maintained a living affirmation of human-nature entanglement, storying, and response-ability for many thousands of years (Danowski and Viveiros de Castro, 2017; Povinelli, 2016; TallBear, 2017; Todd, 2016; Whyte, 2018). And while digital technologies and media are currently enabling an increasing proliferation and exchange of apocalyptic imaginaries across cultures, it is important to note that science fiction extinction narratives are by no means universal or even necessarily communicable across situated cultural and geographical contexts.

Potawatomi scholar Kyle Whyte (2018) observes a large degree of "white panic" in science fiction portraying climate crises as dystopian in Western societies, which he argues are not "the same science fiction imaginaries" (p. 226) shared by Indigenous peoples. Rather, Indigenous climate change imaginaries are often enmeshed in spiralling time as continuous narratives with descendants and ancestors where the present has *already been dystopian* for as long as Indigenous peoples have faced the threat of colonial-capitalistic violence. In this respect, climate change is not located in some distant calamitous future, but in a present that Indigenous peoples continuously experience "through transformations of their societies induced by colonial violence" (Whyte, 2018, p. 224). The overwhelming whiteness of apocalyptic climate change narratives arguably perpetuates this violence, often placing Indigenous peoples in historical categorisations where colonisers become the "saviours" through the imposition of techno-scientific logics.

Thinking in conversation with Zoe Todd, Kyle Whyte, and numerous other Indigenous scholars, Demos (2020, np) argues for a speculative, thoughtful, and creative engagement with multiple and complex "world-ending events, as a progression of ruins that informs present anxieties and realities ... indissociable from the current climate breakdown". Resisting the End of the World narrative as a universalising trope, Demos argues for a more subtle and focused attention on how different worlds are ending, and have already ended, in a variety of different ways.

> Current fears of the world's end are ... importantly contextualised by Indigenous voices that view them as a mode of settler anxiety, haunted by those centuries-old histories of colonial violence, climate-changing brutality, and genocide-directed militarism- forces that have long disrupted fragile ecologies integral to native lifeworlds and continue to do so in the present.
>
> *(Demos, 2020, np)*

An end of *a* world, in this sense, describes the ending of a lifeworld and a way of life for a people, a place, a species, an ecology. Indigenous, Black, Latinx, Asian, and many other decolonial studies emphasise both the geographically situated and bodied repercussions that arise "in the wake" of a world's ending (King, 2019; Sharpe, 2016), as well as the anarchic political powers of the speculative in sustaining the ongoingness of a particular lifeway (Keeling, 2019). Alexis Pauline Gumbs (2015), for instance, describes her book *MArchive: After the end of the world* as a speculative documentary, written in collaboration with "the survivors, the far-into-the-future witnesses to the realities we are making possible or impossible with our current apocalypse" (p. xi). The book is written in speculative and poetic prose from the perspective of future "black feminist metaphysicians" sifting through artefacts after the extinction (or unrecognisable transmutation) of the human as both a species and a category of being. Rather than a single ending to a single world, her future "post-scientists" find multiple worlds ending in a variety of ways depending on the elemental conditions and transmutations of life in a bid for survival.

Minor Gestures and Collective Study

As a way of experimenting creatively with the emergent structure of alternative possibility spaces, these examples demonstrate how speculative fiction offers an ontological tool for conducting social experiments that do not seek to transcend or suspend the political, but rather shift the register of sociality into the sphere of micropolitics and minoritarian resistance (Manning, 2016; Rolnik, 2017; Shaviro, 2015). In their book on Kafka, Deleuze and Guattari (1986) describe the workings of a "minor literature" operating in the cracks, interstices, and "minor keys" between the majoritarian, State-controlled regimes of Art, Literature, Science, Technology, Philosophy, and Education. A minor literature is not concerned with the personal narratives of individual characters, but with the social and political values and speculative collectivities that are mobilised in and through the work (p. 17). Later, in *A Thousand Plateaus*, Deleuze and Guattari (1987) discuss the aberrant becomings of minor languages that introduce continuous lines of variation into majoritarian semiotic regimes and modes of social reproduction.

In her book *The Minor Gesture*, Erin Manning (2016) further conceptualises the minor as a *gestural force* that draws out the potential variations of events in ways that move outside of normative standards and classifications.

> The minor is a gesture insofar as it punctuates the in-act, leading the event elsewhere than toward a governant fixity of the major, be it the major in the name of normative political structures, of institutional life, of able-bodiedness, of gender conformity, of racial segregation.
>
> *(p. 7)*

When we asked the CC+Me researchers how they thought climate change would affect their lives ten years in the future, we were surprised to receive responses that offered minor variations on the intensive dimensions of everyday life.

> I pulled myself out of bed for the first time in days, walking tiredly towards the mirror that hung on my wall. Even through my pajamas I could count my ribs easily. As if on cue, my stomach growled and ached. I hadn't eaten in days. I couldn't afford to.
>
> **Lara, age 11, CC+Me Researcher**

Lara's speculative response operates in a minor key to the extent that it gestures toward the intensive qualities of an undesirable future, and yet a future that is all too common in the present. Lara's image of a near-future ravaged by climate change is not grandiose. Rather than the monumental apocalypse of human extinction it invokes the tiny, everyday apocalypse of hunger and mental illness. In doing so, her words cut open a "potential variation ...from within experience itself, activating a shift in tone, a difference in quality" (Manning, 2016, p. 1). As Manning notes, the minor gesture is both speculative and affirmative while refusing to "ignore the pull of the tragic" (p. 6). Like listening to a musical dirge played in a minor key, Lara's words invoke the "tragic joy of a dramatisation that affirms difference" (p. 220).

It is important to note that Lara's response was not composed in isolation, but as part of an intergenerational research collective working towards a collaborative series of public events and exhibitions. Manning (2016) emphasises that the minor always involves collective modes of attunement to movements that introduce variation into everyday modalities of social life, whether through the play of language, movement, art, or any other practice that shifts the tonality of events to "invent new modes of life-living" (p. 8). Drawing on Harney and Moten's concepts of "study" and the "undercommons" in the radical black tradition of activist thought, Manning affirms that the minor gesture is a collectively generated variation that arises at the nexus of thinking and making: "it is study, it is practice, it is speculation" (p. 13).

In cultivating speculative fiction as a collective act of study that is open to flux, difference, and the continuous variation of experience, we hope to develop its capacity to function as a critical and creative mode of educational practice in the construction of alternate climate child imaginaries. As noted in Chapter 1, and developed throughout this book, our turn to minor modalities of practice and inquiry reflects an engagement with micropolitics that is inseparable from the more-than-human affects and vital drives that are the Earthly preconditions for life (Guattari, 2008; Rolnik, 2017). Our work in the CC+Me project contextualises speculative fiction as a collective mode of creative experimentation and micropolitical resistance that is in conversation with a geophilosophy of the Earth and its immanent relation with thought.

Geotrauma and Fabulation

> I think that leaving Earth to "breathe fresh air" is like breaking something and running away. We wrecked our Earth and now we move on - possibly to wreck another.
>
> **Finn, age 12 years, CC+Me Researcher**

While young people's engagement with speculative fiction in the CC+Me project reflected an intensive attunement to minor languages and micropolitics, the techniques of fabulation that they developed were consistently orientated towards geophilosophical concerns operating at planetary scale. As originally introduced by Nietzsche and further developed in the works of Deleuze and Guattari (1987, 1994), geophilosophy situates the acts of thinking and living in direct relation with the movements and compositions of Earthly processes. "Thinking is neither a line drawn between subject and object nor a revolving of one around another. Rather, thinking takes place in the relationship of territory and the earth" (Deleuze and Guattari, 1994, p. 85).

Geophilosophy establishes the Earth as a metaphysical surface for thought through the formation and deformation of territories, situating the speculative faculties of imagining and fabulation within an ecology of Earthly powers. Territorialisation refers to the expressive process of marking out the conceptual, semiotic, social, and physical milieus that afford places for co-habitation and exchange, along with the associated capture of earthly environments and milieus (Deleuze and Guattari, 1987, p. 311). Alternatively, deterritorialisation indexes the dissolution or abandonment of existing territories in order to form new assemblages through the ongoing variation of thought, vectorised movement, de- and recomposition, and other modes of nomadic experience. Where territorialisation harnesses the tendencies towards organisation, boundary-making, codification, structure, stability, habit, and finitude, deterritorialisation releases the forces of chaos, disorder, variation, mobility, and infinitude. These differential forces function in iterative cycles, such that each territorialisation establishes the conditions for possible deterritorialisations, which in turn establish the potentials for reterritorialisation under an altered set of ontological conditions or "multiplicities". Geophilosophy thus reimagines the activity of thought as an environ/mental process in continuous contact with fluxes in the Earth itself, "as more of a movement of multiplicities that pertains to territories rather than a cognitive faculty restricted to already formed subjects" (Parrika, 2015, p. 21).

In *What Is Philosophy?*, Deleuze and Guattari (1994, p. 89) describe how the Earth occupies a zone of relative and absolute deterritorialisation. Where relative deterritorialisation describes "the historical relationship of the earth with the territories that shape it and pass away", absolute deterritorialisation refers to the Earth's continuous passage "into the pure plane of immanence … of a Nature-thought of infinite diagrammatic movements" (p. 89). The Earth, in this sense, provides a deterritorialised metaphysical surface on which territorial compositions come and go relative to historical states of affairs, while also serving as an absolutely deterritorialised plane for a Nature-thought that carries the virtual potentials for "the creation of a new future earth" (p. 89). Geophilosophy is thus intimately connected with speculative techniques of fabulation and the creation of alternative worlds in response to geotraumas continuously inflicted on the Earth by colonial-capitalist regimes.

> If geophilosophy is a philosophy that grasps thought in relation to earth and territory, then it is a philosophy that, perhaps unconsciously, grasps thought in relation to two traumas, one precipitated by the accretion of the earth and the other ensued by the determination of the territory. Whilst the former trauma lies in the consolidation of the earth as a planetary ark for terrestrial life against the cosmic backdrop, the latter is brought about by a combined geographic and demographic determination of a territory against the exteriority of the terrestrial plane and fluxes of populations of all kinds.
>
> *(Negerastani, cited in Woodard, 2013, p. 14)*

Negerastani's articulation of two distinct forms of geo-trauma is helpful, as they distinguish between the plundering, sedimentation, ungrounding, and stratification of the territories to sustain unregulated economic growth, and the social control of geographic territories that ensures dependency on petro-capitalism into the future. The first geo-trauma has the effect of securing the ongoing capture, commodification, and appropriation of the Earth's ecological systems and milieus by the capitalist-colonial regime. Yet the secondary trauma involves the stifling of minor gestures and collective acts of study that might remediate environmental catastrophe by generating new ecologies of belonging and participation. This form of geo-trauma effectively traps individuals and societies within a present that appears both intolerable and insurmountable, such that we "continue to undergo shameful compromises" rather than exploring creative alternatives for future ways of life (Deleuze and Guattari, 1994, p. 108).

It is in this sense that speculative fabulation and geophilosophy converge as modes of "resistance to the present" capable of summoning forth "a new earth [and] a new people" which are lacking in the current world (Deleuze and Guattari, 1994, p. 99). Fabulation thus involves a summoning of what Deleuze (2002, p. 185) termed the "power of the false, a quality through which the whole of life and its particularity is affirmed and made active". As Manning (2016, p. 220) explains, fabulations activate powers of the false as "affirmative intercessors into the everyday that come alive at the interstices where narration composes with the ineffable". It is important not to confuse the "new earth" and "new people" of geophilosophical fabulation with Romantic visions of an idealised utopia that somehow transcends ecological devastation and social injustice. While speculative fabulation may offer "a resistance to the present in the hope of a better future", as Bogue (2011, p. 82) notes, "that future cannot be predicted, nor can its superiority to the present be assured". Speculative fiction does not offer a solution to geo-trauma or a prediction of future ideals, but an opening up of possibility spaces for collective thought experiments *in continuous contact with Earthly powers.* When coupled with geophilosophy, speculative fiction becomes an ontological tool for exploring the possibilities of minor logics and geopolitical structures that take relational co-existence, creativity, and difference as the basis for "the creation of a new future earth" (Deleuze and Guattari, 1994, p. 89).

Becoming Aberrant

In the Climate Change and Me project, children and young people used speculative fiction as a creative research method for experimenting with alternative climate futures and imaginings of Earthly life. This process was set in motion through the project's open-ended research design and participatory methodology discussed in Chapter 3, which allowed us to work collaboratively with children and young people within a "research playspace" (Cutter-Mackenzie and Rousell, 2014). Initially this involved hosting research workshops in local schools, where children and young people engaged with climate science data, learned to conduct open-ended ethnographic interviews, take field notes, and use creative practices such as drawing, photography, and video as creative techniques of inquiry. We also introduced them to the customised and secure social media interface we had designed to enable the young researchers to post and comment on each other's findings, as well as initiate their own small projects, discussions and games within a blogspace called the "CC+Me Hangout" (see Figure 6.1). The children and young people were encouraged to take on fictional avatars and alternative identities within this online research space, and we also suggested that collaborative thought experiments could be used to explore different climate change futures and social responses.

Jasmyne joined our research workshops as a seventh grade student in a local high school. She rarely spoke during the workshops, but we noticed her continuously sketching in her research notebook or typing furiously on the iPads that we had offered them to use. Several days after the research workshops finished, a series of unusual blogposts began to appear in the online CC+Me Hangout space (see Figure 6.1). The posts were uploaded under the name "Aoihikidashi", or "Aoi" for short, and were often accompanied by detailed Manga-style drawings. It turned out that Jasmyne was using Aoi as an avatar for developing a speculative series of characters and worlds that grew out of our research workshops and online platform. The first appearance of Aoi on November 17, 2014 marked the beginning of an ongoing series of dated entries on the project website, as Jasmyne outlined the contours of an alternate world in which children develop mutant forces and bodily augmentations that enable them to resist and adapt to climate change. Jasmyne's fictional world drew on her deep immersion in the aesthetic and narrative conventions of Manga culture and related genres of Asian speculative fiction. And yet the characterisations and environments she conjured also touched on elements of climate science, environmental activism, anarchic politics, biogenetic mutation, magic, and digital life that blurred distinctions between cultural semiotics and grammars.

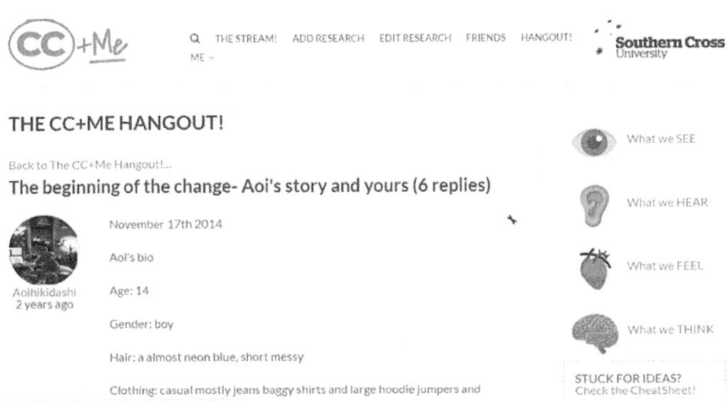

FIGURE 6.1 Screenshot from "The CC+Me Hangout" showing Jasmyne's first entry introducing her avatar character Aoihikidashi, or Aoi for short.

Within Jasmyne's fictional world, children experiencing these transmutations are labelled "Aberrants" as opposed to "Normals", and hunted down by cyborgs capable of scanning and annihilating those with anomalous genetic mutations. Aoi remains a central character throughout Jasmyne's story, who discovers after a near-death incident that anything he draws in his sketchbook can come alive for a limited time. As a research collective, we witnessed the emergence of this speculative narrative as it continued to evolve across multiple timescales in tandem with the events of our actual research workshops. The timeframe of Jasmyne's entries begins in November 2014, the date of our initial research training workshop, and follows a non-linear adventure of characterisations and transformations of the earthly environment into December 2047. It is through Jasmyne's meticulous extension of characterisation, timescale, and geographical location that her fictional world becomes palpable as creative research, while simultaneously addressing critical posthumanist concerns: the mutations of new forms of life that exceed the human; the erasures of memory through geo-traumatic psychosis; the arbitrary destruction of human and nonhuman entities; the (un)desired affordances and responsibilities of aberrant neurodiversity; the rapid proliferation of nano-technologies; and the spectral hauntings of cyborg politics.

Jasmyne's creative research initiative also offered an opportunity for ongoing collaborations with other young people and co-author David, who took on fictional avatars and posted their entries into the Hangout space on a weekly basis. Over time, the entries developed into a non-linear collaborative narrative in which the voices of individual authors became entangled and, at some points, indistinguishable. These entries were then shaped into an illustrated short story called "The Beginning of the Changes" in the co-edited book of children and young people's research that accompanied the travelling *Past Now Future* exhibition (Rousell and Cutter-Mackenzie, 2015). The following section offers a series of abridged entries from "The Beginning of the Changes", as illustrated with Jasmyne's characterisations and sketches. The entries are presented in the order they were submitted to sustain the non-linear movement between fictional voices, places, and timescales that emerged through the collaborative process.

Entry 1: November 17, 2014

Name: Aoi
Age: 14
Gender: Boy
Hair: An almost neon blue, short, messy
Clothing: Casual, mostly jeans, baggy shirts and large hoodie jumpers and Large sneakers
Accessories: One silver earring and a gold family ring

FIGURE 6.2

A room that looked like a bomb hit it, clothes scattered messily on the floor. A bed that didn't look much like a bed, more like a table with sketchbooks scattered on the end. A bag crumbled in the corner with a clatter of key chains and bulging with the shape of a lunch box. This was Aoi's room. He was messy, though he would claim this as organised chaos and leave it at that. It wasn't that he couldn't clean, no he was perfectly capable of that. It was just that he couldn't be bothered, or he would claim that he had "bigger problems to focus on", though that was a lie. Even though Aoi was bright all he ever focused on was drawing in his sketch book. His mother and father worried for him, but what could they do? It was his life and they couldn't force him to do anything. They loved him so much but there was no point to try. He just wouldn't go out of his way to please them if that meant giving up his drawing.

Though Aoi was a loving son and thought the world of his parents he didn't want to worry about getting good grades just to please them alone. Even so, if his grades dropped below average or just on the average line he'd study hardcore to get back his place. Though it seemed as if he didn't care, he actually did. At least enough to work hard to get above average again. It wasn't the fact that Aoi didn't care about school, he just cared about his artwork more. He decided that when it came to university entrance exams he would work harder at school than he had before, but until then the topics discussed at school just weren't engaging for him. The only subjects he actually liked were science and art.

Aoi was interested in the workings of the universe and was a full blown realist. He had always cared about the earth's geological problems and biological workings. He was always interested in that, but there were so many branches of science that he was overwhelmed with choices. But he had always been sure that he wanted to help people. He considered going into environmental science. But when he thought about it, helping scientists in the lab might be difficult for someone who just loved to draw so much.

Then when he thought of giving up on science and becoming some generic worker for a large company he found out about climate change. It was through a university talk that he found out about it in depth. He knew it affected the climate and had something to do with the Ozone layer but he knew nothing else about the topic. But after the talk he decided to search for some information and he took to the topic of climate change like a duck to water. He was so strongly taken by the topic he began to post about it on social media. Aoi was so interested in this topic and he wanted people to become more aware but how could he make a difference?

He was only a child in the eyes of the public, not likely to be taken seriously. What could *he* do to help stop climate change?

Entry II. November 18, 2047

Name: Elbereth
Age: 47
Gender: Woman
Hair: White with slight streaks of blue
Clothing: Old clothes with holes that have been eaten by moths
Accessories: Two pearl earrings and a fox mask that she keeps in a box under her bed
Companions: A wise and tired old dog named Turin

FIGURE 6.3

The wind blows in the door again and I can hear the rush of the water breaking on the cliffs below. Turin barely moves at all, he just opens one eye and looks at me as if I've created the disturbance. The air is full of frozen salt crystals, they bite my cheek as I brace the door closed with the last of my heavy books. Soon I'll have used up all my esoteric books for fire-starters and everything in them will probably be forgotten. It's been too long since I've felt the power of the earth in my fingertips, the electric current that turns the waves away from the valley below. Now I lack the strength, and the memories flow back like the tides that nobody can stop from washing in.

But what's that? Turin is up, stretching and starting to growl at the same time. I hope it's not those raiders again, last time they cleared me out of rice and oats faster than I could get out of my rocking chair. Something is scratching at the wooden door. I grab the knife from next to the cutting board where I'd been chopping kale from the garden. It's only a bread knife, but it's better than nothing.

"Who's there!" I say in my deepest, scariest voice. I open the door just a crack, and can just make out the shadow of a person against the pelting wind and rain. It's a child, no a man, with a face that looks so familiar and yet cloaked in shadow. I open the door all the way and he falls into my cabin, drenched and exhausted on the floor.

"Who are you?" I say sternly with the kale knife gripped steadily in my hand.

"Aoi," the voice faintly replies.

FIGURE 6.4

Entry III. December 5, 2014

All the kids at high school had been invited to the research workshop about climate change. Aoi found himself doing a practice interview with a talkative girl named Elbereth. "I think the Earth has already transitioned into a new geological era," she was saying. "Humans have become like this viral force that's affecting every ecosystem on the planet."

That afternoon after school they went for a walk in the old forest behind the school, Elbereth with her notebook writing stories and Aoi with his sketchbook drawing the creatures that sprang into his mind. They came to a rushing stream that had swollen with water after the recent flood. Aoi started to make his way across the stream on the wet log. Suddenly, he slipped on a patch of moss and fell into the rushing water. His head hit a rock and his body was swept downstream.

Elbereth ran along the bank as fast as she could, but when she finally caught up to him his body was cold and lifeless. Desperate, she placed her hands on his heart and summoned all the power in her body to put warmth into him. She felt a vital energy enter her body from the forest floor, pulsing through her arteries like blood, producing heat that made her skin burn as if touching flames. Suddenly Aoi sputtered and sat up, looking at her with wild eyes.

Later they found Aoi's sketchbook soaked and lying upside down alongside the rushing stream. It was moving slightly. Aoi picked up the sketchbook, and jumped back in surprise- underneath it was a frog-like creature with pale translucent wings. It jumped onto the nearest rock, then flapped its wings and flew away.

With trembling hands, Aoi turned over his notebook to the page he had been working on before he fell into the stream. There on the page was the very creature they had just seen fly away- an amphibian with the wings of a dragonfly. Aoi looked at his drawing hand in amazement.

The flesh had changed to a strange blue colour, and the nails had become like black claws.

FIGURE 6.5

FIGURE 6.6

Entry IV. March 23, 2021

Senior detective Shinohara called a meeting of the task force.

Everyone stood silently in the centre of the task force building. Nerves were high. It was rare that the entire task force would be asked to meet unscheduled. Everyone was tense, even senior detective Shinohara looked tense. The eerie silence continued for another few seconds until Shinohara broke it.

"Task force, we are gathered here today to see the latest technology from Astral Labs to combat the ever- growing aberrant infestation. I will now hand the floor over to Professor Arima"

A man emerged from the other side of the room. He wore a white coat and carried with him a brief case. He stood next to Shinohara in the centre.

"I am professor Arima head scientist at Astral Labs, I will attempt to keep this explanation brief." He placed down the brief case.

"This," he said pointing to the brief case, "is the future of aberrant hunting. It will allow you to find them and neutralise them faster than ever before."

"Is this some kind of sick joke?" junior detective Yakamo interrupted. "How on earth is a brief case going to help exterminate aberrants?"

"Yakamo, keep quiet," Shinohara ordered.

"Don't worry detective, questions were bound to be asked," Arima said kindly.

"Here, allow me to demonstrate." He pushed down on the handle of the brief case. The case split in two revealing a block of metal. The block then started to transform, slowly taking humanoid form. The task force was amazed. Once it had completed its transformation Professor Arima continued with his explanation.

> This is Amadron, the latest in aberrant hunting technology. This is military funded weaponry with a built in psycho-pass able to scan to a fifty meter radius. It is fully equipped with stun guns but for your purpose, of hunting criminal aberrants, we also equipped him with a kill mode that can be activated with the removal of a chip in the chest cavity. Made of carbon steel it is a light but hard metal that can easily be transported in a brief case such as this. It also has a specific targeting system which allows you to hunt a certain aberrant; all that is required is a sample of blood. With on-board facial recognition system and a fully equipped medical kit it is by far the most powerful weapon for aberrant extermination ever invented.

Entry V. April 15, 2021

Aoi did not want to go to the lecture.

"Come on Aoi," Elbereth moaned as she retrieved her jacket from its lowly hook. Aoi sat stubbornly on the couch pretending he couldn't hear her groaning. Elbereth was now becoming impatient. "Aoi it's just one lecture, would it kill you to come along?"

"Yes, if I have to sit there listening to another pseudo-scientist babble on about a non-existent creator then yes, it will mentally kill me."

FIGURE 6.7

Elbereth threw her coat over Aoi who was still firmly seated on the couch. "Come on, I listen to all of your scientific lectures."

"That's different and you know it," Aoi mumbled underneath the coat. Elbereth walked over and removed her coat from Aoi's head. She sat next to Aoi and sighed heavily.

"Just do it Aoi. For me," she said clasping her small hand over his larger one.

"No," he shortly replied removing her hand from his. "It's too dangerous for us to go out. We need to stay home where we're safe."

"No Aoi, we can't lock ourselves away in this apartment forever." Elbereth lifted herself from the couch and retrieved her purse from the hook next to where the coat hung. Aoi hoisted himself from the couch and confronted her before she could leave.

"You're mad at me," he said sadly. He wrapped his arms around her and lightly hugged her protectively.

"Aoi." She attempting to push him away. "I don't want to be caged like an animal, I know we are different but is it really that bad?" Aoi stood silently.

"I guess it isn't that bad," he replied. He lightly rubbed the back of his head. Elbereth saw his slight rub and knew immediately that he was hiding something.

FIGURE 6.8

"Are you lying Aoi?"

"No, I'm not" He rubbed the back of his head again.

"You are, you were rubbing the back of your head!" Elbereth said angrily.

"No, I wasn't," he replied while rubbing his head again.

"Aoi you are a terrible liar, you rub the back of your head whenever you're hiding something."

"Yeah, I do," Aoi said smiling weakly.

"What happened?" Elbereth asked concerned.

"They almost caught me," he replied.

"Who?"

"The task force. They caught me off guard with their psycho-passers. They must have just recently upgraded them," he explained.

"Aoi tell me exactly what happened," she said, taking off her coat and placing down her hand bag to sit down on the couch. Aoi walked over and sat gently next to her.

"I had just come from the shops with our groceries," Aoi began. "Across the street I saw them using the psycho-passes. I thought nothing of it." He paused because he could see Elbereth's dread.

"All I did for a safety precaution was put up my hoodie. Thank God I did," he laughed. "The psycho-pass beeped as soon as I walked by. I saw the officers coming for me so I panicked and ran. Then everyone knew that I was an aberrant. The innocent don't run, so they chased me. I had my note book on me though so I quickly took out the mask …"

"Oh god Aoi. Why would you do that?" Elbereth interrupted. "You know what they associate that mask with. Now they know we're here, we're going to get caught again and questioned." Elbereth shuddered. She brought her hands to her shoulders. They began to heat up. Her skin was burning. Aoi brought his hand up to hers and unclasped them from her shoulders. They began to cool down as soon as he touched them.

"Don't be scared. It's alright, I can protect us," Aoi said reassuringly. But Elbereth had a hard time believing him. "Trust me," he begged.

Elbereth looked at her hands. "I do trust you Aoi, it's just ..." her voice trailed off.

"Just what?"

"If you're protecting us who will protect you?" She looked scared. She pulled herself away from Aoi, picking up her coat and placing it over her shoulders. Then she retrieved her handbag and left, slamming the door behind her and leaving Aoi alone in silence.

Aoi sat there for a moment contemplating the recent string of events. He looked to the floor where he saw his notebook face down. He gently lifted it off the ground looking sadly at the page it was open on. The drawing of his mask. He tugged lightly at first, then ripped it out completely, angrily scrunching it up, throwing it to the floor. He threw aside the note book and went to the coat rack retrieving his coat. He turned off the apartments light then began chasing after Elbereth.

As the door closed submerging the room in darkness, the paper began to slowly unfold revealing a scrunched up drawing of a yellow smiley face.

* * *

FIGURE 6.9

Elbereth buried herself into her coat.

It had been a long time since she had actually left the apartment without Aoi's company. He was overly protective and always insisted on accompanying her for her so-called protection. In the end she knew that it wasn't the only reason he accompanied her, the real reason was that he was afraid of being alone. So she put up with it. No real harm could come of him following her to the grocery store.

She walked steadily looking at the graffiti scribbled across the side of the buildings. Many were rude words or artworks so she paid no attention to the messages. She soon came across the task force headquarters where wanted posters of criminals were pasted along the side of the discoloured brick walls. But criminals weren't the only ones with wanted posters.

FIGURE 6.10

The wall had a section for "aberrants". This was the dim name that the humans had classified those with powers under. She walked over to the wall and looked for her wanted poster. It didn't take long to find a poster of her with her mask on. It was labelled "Firefox". Elbereth could feel heat emanating from her hands. *Dangerous, volatile, and will kill without remorse?* They were all lies!

She grasped a corner of the poster and tore it down from the wall, using the heat emanating from her palm to set the poster ablaze with fire.

Entry VI. November 20, 2047

I wake in my rocking chair to the smell of oatmeal cooking over the fire. No, it wasn't all a dream. My oldest friend Aoi is crouching over the flames, stirring the pot with a long wooden spoon. He turns and looks at me, the piercing clarity returning to his eyes by the second.

"Aoi," I say. "I can't believe you're here. How did you find me after all those years?"

"There was just this map, Elbereth, a map in my mind. And I followed it to your door. But tell me, please. I think I've lost my memory for thirty years. Are we the only ones left in the world? Did my visions actually come true?"

"Many came true, and we are deep in the Changes as they have spread across the Earth," I explain.

The power grid shut down five years ago, and the Netscape was destroyed not long after. Wireless mesh networks still communicated for a while- I have a node here that kept me updated with the latest information until it rusted out from the saltspray last year. Transmissions were coming through from communities up and down the coast that had formed their own collectives, rebuilding their towns and villages to adapt to the Changes. The cities are apparently lawless and mostly deserted. People like us have been helping- using our earth powers to protect and heal the communities- and fighting off the raiders when we need to. Do you still have the power that you did when we were young?

"It was like I'd forgotten it all when the shadow overtook me in the forest. I forgot how to draw, and couldn't summon the focus to learn again."

"Here, Aoi. Take this charcoal from the fire. See if it can work again."

Aoi takes the charcoal in his trembling hand. I can't believe he could have been cursed and lost his powers for so long. He begins to scratch out forms on the floor of the cabin. The black outline of a bird takes shape, Aoi is working more quickly, gaining confidence. The bird starts to move, just slightly at first... and then it lurches off the floor, it's like a baby just learning to fly, dipping up and down, crashing into the wall, picking itself back up. And then it's flying smoothly around the cabin, a black bird made of a smoky charcoal substance, and it comes to land on Aoi's hand. The bird turns from charcoal to a gleaming gold. Suddenly Aoi looks just like his old self.

"It's back, I can't believe it, it's back!" he says with gleaming eyes. "Elbereth, I don't know how but somehow you've healed me from the curse. We have to get out there and help."

"But my powers have dwindled too, Aoi," I say. "I don't know what happened, but they just faded over time."

Aoi's bird flies over and lands on my outstretched hand. It starts pecking at the lines and creases in my palms. Suddenly I can feel the power flowing back in, from my heart all the way into my hands. Tentatively, I point my palms at the fire and concentrate with all my might. Fssssss ... the fire goes out with a sizzle. I squeeze my eyes shut, putting everything into my hands. Crackkkkk.... The smoldering wood of the fire is crystalising into ice. It's working again, my powers are coming back to me. "Aio", I say softly. "We're us again."

FIGURE 6.11

Entry VII. December 1, 2047

"Aoi, the food's ready".

"I'll be out as soon I can. Just have to finish this," I say softly, trying not to show Elbereth how desperate I am to finish the drawing. Drawing is the cure to my illness. When I have a sketch book near me I feel my power and memories rushing back. Drawing makes me feel invincible. It's like I draw power from the image, and the better the image the better I feel.

Then we hear a ghostly rapping on the door. I freeze, while Turin barks furiously at the door. "Who's that?" Elbereth calls from the kitchen. I can hear her kale knife coming down with force onto the chopping board.

"Come out *meat*," a voice calls tonelessly. I feel my heart skip a beat. I've heard that voice before. Then the door breaks open and an arm of steel emerges through the hole in the wood. The hand draws back and a pulsing robotic eye peers through. "I found you Anthropos!" the toneless, mechanical voice calls out.

I bolt up, sweat pouring down my face. "We have to leave," I mutter nervously, my voice cracking like glass as I speak.

Elbereth just stares. "Who is that Aoi?" The metal hand is scraping furiously at the wood like a caged animal.

"It's Shudon. He's finally found us." I push Elbereth forward towards the back door, casting a quick glance back. Shudon is crawling in and rapidly coming towards us. I know I won't make it, but Elbereth has a chance. So with one great shove I throw her out the back door of the house and scramble for my sketchbook. He's almost here. I begin to flip through the book, desperately searching for the right page. Then I feel a lack of air and cold metal fingers wrapping tightly around my neck choking me. I drop my sketch book, it hits the floor with a loud thud. The steel fingers tightly curl around my neck. I can feel the air getting stuck in my lungs, the sharp feeling of suffocation. I move my mouth but only a faint whine comes out, like the sound of a whale.

Setting the Coordinates for Experimentation

While Jasmyne's speculative narrative continues to splinter across non-linear storylines that introduce a range of other characters and situations, we have decided to leave the story of Aoi and Elbereth on a cliff hanger in order to foreground the affective and minor movements that unfold over the course of these entries. While the entries above offer numerous possibilities for analytic interpretation and educational application, here we prioritise the minor gesture of speculative fiction as an ontological tool for re-imagining climate change education through the aesthetic production of "quality as difference" (Manning, 2016, p. 220). In adopting the theoretical imaginary of geophilosophy for this purpose, our analysis focuses on the capacity for speculative fiction to deterritorialise the concepts and practices of climate change education through the production of new territorial assemblages as functions of creative expression. Deterritorialisation, in this geophilosophical sense, occurs when "expressiveness no longer functions strictly for the territory but expresses a becoming … a new, unforeseen territorial assemblage" (Young, 2013, p. 307).

Our discussion of this work begins by acknowledging that the emergent, existential territories generated through Jasmyne's entries are not generic or universal but "contingent and specific", with particular nodes of inception and sets of topological coordinates that condition the structure of the social worlds it produces (Bogue, 2011, p. 82). More specifically, we see the entries operating as an open constellation or topological series that de- and reterritorialises the actual events of the Climate Change and Me research, including the research workshops, the collaborative fieldwork, the online social media space of "the Hangout", and the ThinkTank analysis workshops where other children began to write their own characters into the work. In short, the entries above do not simply "float into the world from nowhere" (Whitehead, 1978, p. 244). Rather, they are intimately connected with the spatial and temporal coordinates of the research itself as a series of events that took place within dated places and times, and which continue to ripple out in both empirical and speculative dimensions.

This speculative empirical proposition allows us to understand fabulation as a process that generates *new* entry points or nodes in a series of events that are always already in motion. By introducing new nodes of inception into a series of events, Jasmyne's narrative sets out "protocols of experimentation [which] provide a certain scaffolding for resistance" through the collective act of creative fabulation (Bogue, 2011, p. 82). As Deleuze and Guattari (1994, p. 110) suggest, the act of fabulation is an act of political refusal that resists the constraints and injustices of the present. Indeed, it is this very resistance to the material and discursive constraints of the research itself that enabled Jasmyne's visual and textual entries to become "reference points for an experiment that exceeds our capacities to foresee" (Deleuze and Parnet, cited in Bogue, 2011, p. 83).

In this way, we see the serialisation of Jasmyne's narrative experiment extending beyond what is immediately visible, intelligible, or recognisable as the present, inviting co-authors and readers to pass seamlessly through different spatio-temporal coordinates in relation to geo-traumatic events both actual and virtual, real and imagined. This is the work of the minor. "There are dimensions here", Deleuze (1990, p. 128) writes, "times and places, glacial or torrid zones never moderated, an entire exotic geography which characterises a mode of thought, as well as a style of life". Rather than operating as a descriptive account, depiction, or analogy of alternative futures, Jasmyne's speculative narrative *simulates* a possible world of sensation, affectivity, and thought that is transversally related to the empirical dimensions of our present lives.

The Beings of Fiction

It is through this experimentation with possible futures that Jasmyne's entries enable the geo-traumatic effects of climate change to become palpable through the experiences of Aio and Elbereth, as what we have called "children of an Earth to come" (Rousell et al., 2017). Both Aoi and Elbereth are examples of what Bruno Latour (2013, p. 251) calls "beings of fiction", conceptual personae or "actants" who are "made of the same materials, the same figures" as the apparatuses of scientific inquiry, or the conceptual schematics of philosophy. Latour argues that such fictional characters exert an influence beyond the worlds of the imagination to reveal the speculative and compositional nature of science, technology, medicine, law, and every other practical mode of existence (p. 249). "How can we speak about remote galaxies, particles of matter, upheavals of mountains, valleys, viruses, DNA or [climate change]", Latour asks, "without having at our disposal *characters* apt to *undergo* such adventures?" (p. 250, emphasis in original). In the entries above, our two characters take us "elsewhere, into another space, another time, another figure or character or atmosphere" (p. 246), highlighting the internal and external consistencies of new aesthetic milieus that are also intimately connected with the concrete challenges of our current existence. Yet Aoi and Elbereth are not pure, ideal, or transcendent characters who serve a moralistic or instrumental function. Rather, they are aberrants, mutants, members of "an oppressed, bastard, lower, anarchical, nomadic and irremediably minor race" (Deleuze and Guattari, 1994, p. 109). Despite Aoi's visionary capacity to understand the "earth's biological and geological workings", as Jasmyne writes, he remains "only a child in the eyes of the public, not likely to be taken seriously".

The minoritarian status of Jasmyne's characters is emblematic of the broader marginalisation of children and young people in today's political economies as "minors", in which they are accorded minimal agency on political issues that have a direct bearing on their lives. As minors, Aoi and Elbereth are very much aligned with the "missing peoples" whom Braidotti (2019) sees emerging from within the posthuman condition. For Braidotti, the missing peoples include those "real-life subjects whose knowledge never made it into any of the official cartographies" but also "the virtual ones" that constitute a "complex singularity, expressing the embedded, embodied, relational and affective forces that generate patterns of becoming, of minor science, of intensive shifts" (p. 51). Much like Elbereth and Aoi, children and young people who wish to enact their political beliefs on climate change have very little choice but to become "aberrants" themselves. Moreover, such aberrancy is described in Jasmyne's entries as an "infestation" which must be "exterminated" by means of biometric surveillance technologies such as facial recognition, psychological scanning, and biogenetic identification. This is very similar to the current pathalogisation of youth climate politics by neoliberal governments and educational leaders in the wake of the global school strikes.

Geomorphic Mutations and the Decaying Earth

As beings of fiction, it is Aoi and Elbereth's minoritarian resistance to the normative violence of an intolerable present that constitutes the source of their power and intensity, as expressed through their openness to the chaosmic forces of the Earth under the conditions of geo-trauma. Both Aoi and Elbereth are transfigured by the geological forces of creativity unleashed from the Earth itself, which deterritorialise and intensify their biological structures and bodily capacities to respond to climate change. We suggest that these can be read as *geomorphic* rather than anthropomorphic mutations, in the sense that the characters transform in "rhizomatic connection with the earth (the forces of territorialisation and deterritorialisation)" (Bonta and Protevi, 2004, p. 92). The characters' increased bodily capacitations and intensities are not deliberate, intentional, or filiative in any *human* sense, nor are they the result of a linear, hereditary, or genealogical evolution towards greater functionality, mastery, or well-being. Rather, these mutations can be understood geophilosophically in terms of a dark, machinic, and creative *involution* (Deleuze and Guattari, 1987, p. 238), as an attunement to "planetary decodings" that release becomings which are excessive, unstable, and contagious in nature (Wallin, 2015, p. 139). While Aoi's drawing hand is afforded the capacity to bring images to life that resist climate change, it also becomes monstrous and infected by his sensitivity to the traumas being inflicted on the Earth.

As these elaborate fictional worlds emerged from the midst of our project, it became apparent that children and young people were constructing a climate child imaginary through practices of speculative thinking and fabulation that departed radically from typical Western narratives of climate apocalypse. While there was a geopoetic darkness to these narratives, it wasn't necessarily the kind of darkness associated with the flurry of apocalyptic representations of climate change in popular media that Colebrook (2014, 2019) cites in her recent work. Rather, their turn to the speculative carried a dark aesthetic entangled with experimental practices of knowing, feeling, and caring beyond the human, of grappling with ontological re-imaginings of nature, humanity, and what we understand to be real. This discordant aesthetic accepts climate change as a challenge to think the world differently: darkly, speculatively. Haraway (2016) has described this kind of dark entanglement in terms of an intricate and "tentacular" engagement with the "other-than-human" or "Cthonic" forces of nature, as expressed in her multifaceted figuration of SF as science fictions, speculative feminisms, string figures, science facts, so far.

As Jasmyne and many other children took up this dark line of ecological thought in the CC+Me project, we found that the Earth was no longer being figured as a source of wellness and abundance but as an occulted, diseased, and ungrounded series of decaying and dysfunctional systems (Negarestani, 2008; Wallin, 2015). In these stories there is a haunting sense of an Earth that is active in its own deconstruction, and in the undoings of the characters who attempt to pursue regenerative trajectories of resistance. We also note here the enormous toll taken on the bodies of young climate activists, such as Greta Thunberg, who appear increasingly skeletal, haunted, and drained in media interviews and public addresses. In the stories shared above, the onset of Aoi's mutant capacitation results almost immediately in a melancholic endarkening of his affective state, followed by three decades of amnesia as he is pursued to the ends of the earth by Shudon, a merciless automaton propelled by the techno-scientific regime. Elbereth's initial optimism and green revolutionary spirit quickly give way to a reclusive life of meagre survival, as her elemental powers dwindle with the onset of loneliness and ennui. Rather than being gifts of earthly connection or transcendence, the characters' mutations affect them more like a curse, an infection, an addiction, or a virus, for which their aberrant creativities and strange alliances offer only temporary cures.

Towards a Viral Climate Change Education

Our geophilosophical analysis of Jasmyne's entries suggests that speculative fiction can enable young people to respond to climate change in ways that catalyse "the emergence of new possibilities for life through resistance to the present" (Bogue, 2011, p. 88). This work offers a distinctly dark and viral approach to climate change education research which finds its expression in the proliferation of experimental practices, alliances, architectures, and spatio-temporal nodes of creative inception. While this approach aims to foster creative modes of resistance to a present that appears both unjust and intolerable, it also maintains "a posthumanist ethical commitment to stay with the occulted life of this planet" without aspiring to transcendence, escapism, or any other "way out" (Wallin, 2015, p. 144).

Looking back on the emergence of these speculative narratives and characterisations and their activations of other children's climate imaginaries, we can identify a specifically *viral* strain of climate change education which offers certain strange and ambivalent openings for practical approaches in the classroom. The stickiness of these stories urges us to consider the ways that global warming, nuclear waste, water contamination, or other ecological catastrophes might enter into the classroom not only as discursive topics, but as material and affective forces which pass through and infect the very structures of bodies (Rousell, 2021). This conception of climate change education as virus or contagion therefore differs from developmental, genealogical, or what Deleuze and Guattari (1987, p. 12) might call "arborescent" modes of teaching and learning predicated on the cognitive development of the individual student. It points to *fabulation*, rather than information, as the key to materialising alternative climate change futures through education.

For our purposes here, the viral potentials of such a climate change education might best be illustrated through the narratives' ongoing infection of the learning experiences of other children who encountered them. As part of the climate change curriculum that we trialled in local schools in 2016, over 150 students read and responded to Jasmyne's narratives in a multiplicity of different ways. Many children even wrote themselves into the story by creating their own "Aberrants" with mutant abilities such as:

- controlling dark matter, time, and darkness;
- appearing completely emotionless in the face of fear;
- changing emotions at will and infecting those around me with my mood;
- shapeshifting into any animal or plant;
- putting thoughts into other peoples' heads;
- enabling people to meet themselves and truly feel others;
- manipulating and controlling media communications; and,
- changing or redirecting matter in the general vicinity, as in the creation of singularities.

We note that these abilities clearly resonate with the affective, stylistic, and material inflections of Jasmyne's original entries. They go beyond typical "superpowers" as represented in popular speculative fiction narratives to propose a dark ecology of powers that lurks in the esoteric thresholds between affective intensity and speculative fabulation. This is evident in the children's focus on the potential transfigurations of both matter and life, as well as an emphasis on the manipulation of affective states as conditioned by the narrative's "dark" ecological aesthetic.

The contagious qualities of these speculative narratives were also reflected in a range of creative responses crafted by students in other schools that trialled the curriculum. Of particular note was the development of an interactive multimedia platform under the rubric of *Atlantic Adaptation Theory* (AAT). As conceived, designed, and managed by an enterprising collective of year-six students using a customised Discord server, the AAT web platform focused on theorising the possibilities for biological adaptations to climate change, including explorations of bio-physical mutation, breath analysis, space exploration, organic terraforming, bioluminescence, and hypnosis, among many other speculative areas of inquiry. Participation in the AAT platform was through a game-based system in which students' avatars advanced in rank (from apprentice, to researcher, to theorist) based on the quality and extent of their research contributions. The contagious influence of Jasmyne's original narratives on the AAT platform can thus be seen in the students' development of a real-time collaborative architecture, along with the platform's focus on bio-hacking, living labs, and bodily mutation rather than techno-scientific "solutions" to environmental crisis. The viral quality of these proliferations across digital platforms has continued to inform our work in subsequent projects, including the design of the Climate Action Adventure! application which we have co-created with young people based on insights from the original project. While impossible to quantify, this contagious proliferation of affective resonances and relays from one emergent node of the research to another ultimately presented one of the most salient and surprising findings of the larger Climate Change and Me project.

References

Bogue, R. (2011). Deleuze and Guattari and the future of politics: Science fiction, protocols and the people to come (Article). *Deleuze Studies, 5*, 77–97. https://doi.org/10.3366/dls.2011.0038.

Bonta, M., & Protevi, J. (2004). *Deleuze and geophilosophy: A guide and glossary*. Edinburgh University Press.

Braidotti, R. (2019). A theoretical framework for the critical posthumanities. *Theory, Culture & Society, 36*(6), 31–61.

Braidotti, R., & Bignall, S. (2019). *Posthuman ecologies: Complexity and process after Deleuze*. Rowman & Littlefield.

Colebrook, C. (2014). *Death of the post human: Essays on extinction, vol. 1*. Open Humanities Press.

Colebrook, C. (2019). A cut in relationality: Art at the end of the world. *Angelaki, 24*(3), 175–195.

Cutter-Mackenzie, A., & Rousell, D. (2014) 'Climate change education in Anthropocene and Holocene times' *Invited symposium presentation at Australian Association for Research in Education (AARE) conference*. Brisbane, Australia.

Danowski, D., & Viveiros de Castro, E. (2017). *The ends of the world* (R. Nunes, Trans.). Polity.

de Freitas, E., & Truman, S. (2020). New empiricisms in the Anthropocene: Thinking with speculative fiction about science and social inquiry. *Qualitative Inquiry*. https://doi.org/1077800420943643.

Deleuze, G. (1990). *The logic of sense* (M. Lester, Trans.). Columbia University Press.

Deleuze, G. (2002). *Nietzsche and philosophy* (J. Tomlinson, Trans.). Columbia University Press.

Deleuze, G., & Guattari, F. (1986). *Kafka: Toward a minor literature* (D. Polan, Trans.). University of Minnesota Press.

Deleuze, G., & Guattari, F. (1987). *A thousand plateaus* (B. Massumi, Trans.). University of Minnesota Press.

Deleuze, G., & Guattari, F. (1994). *What is philosophy?* Columbia University Press.

Demos, T. J. (2020). *Beyond the world's end: Arts of living at the crossing*. Duke University Press.

Doyle, J. (2020). Creative Communication approaches to youth climate engagement: Using speculative fiction and participatory play to facilitate young people's multidimensional engagement with climate change. *International Journal of Communication, 14*(24), 2749–2772.

Ehret, C., & Čiklovan, L. (2020). How speculative designs produce new potentials for education research in digital culture. *Discourse: Studies in the Cultural Politics of Education, 41*(5), 708–722.

Ghosh, A. (2016). *The great derangement: Climate change and the unthinkable*. University of Chicago Press.

Gough, N. (2015). Undoing anthropocentrism in educational inquiry: A Phildickian space odyssey. In N. Snaza, & J. Weaver (Eds.), *Posthumanism and educational research* (pp. 151–166). Routledge.

Guattari, F. (2008). *The three ecologies*. Continuum.

Gumbs, A. P. (2015). *M Archive: After the end of the world*. Duke University Press.
Haraway, D. (2016). *Staying with the trouble: Making kin in the Chthulucene*. Duke University Press.
Harney, S., & Moten, F. (2013). *The Undercommons: Fugitive planning & Black Study*. Minor Compositions.
Jensen, C. B. (2020). Cli-Fi, education, and speculative futures. *Comparative Education Review, 64*(1), 150–152.
Keeling, K. (2019). *Queer times, Black futures*. NYU Press.
King, T. L. (2019). *The black shoals: Offshore formations of black and native studies*. Duke University Press.
Latour, B. (2013). *An inquiry into modes of existence* (C. Porter, Trans.). Harvard University Press.
Manning, E. (2016). *The minor gesture*. Duke University Press.
Morton, T. (2013). *Hyperobjects: Philosophy and ecology after the end of the world*. University of Minnesota Press.
Negarestani, R. (2008). *Cyclonopedia: Complicity with anonymous materials*. re.press.
Nxumalo, F. (2020). Situating indigenous and black childhoods in the anthropocene. In A. Cutter-Mackenzie-Knowles, K. Malone, & E. Barratt Hacking (Eds.), *Research Handbook on Childhoodnature: Assemblages of Childhood and Nature Research* (pp. 535–556). Springer.
Parrika, J. (2015). *A geology of media*. University of Minnesota Press.
Povinelli, E. A. (2016). *Geontologies: A requiem to late liberalism*. Duke University Press.
Rolnik, S. (2017). The spheres of insurrection: Suggestions for combating the pimping of life. *E-flux Journal, 86*, 1–15.
Rousell, D. (2021). *Immersive cartography and post-qualitative inquiry: A speculative adventure into research-creation*. Routledge.
Rousell, D., & Cutter-Mackenzie, A. (Eds.) (2015). *The Changes: Art, research, and writing about climate change by children and young people*. Southern Cross University.
Rousell, D., Cutter-Mackenzie, A., & Foster, J. (2017). Children of an earth to come: speculative fiction, Geophilosophy and climate change education research. *Educational Studies, 53*(6), 654–669. https://doi.org/10.1080/00131946.2017.1369086.
Rousell, D., Wijesinghe, T., Cutter-Mackenzie-Knowles, A., & Osborn, M. (2021). Digital media, political affect, and a youth to come: Rethinking climate change education through Deleuzian dramatisation. *Educational Review*. https://doi.org/10.1080/00131911.2021.1965959.
Sharpe, C. (2016). *In the wake: On Blackness and being*. Duke University Press.
Shaviro, S. (2015). *Discognition*. Repeater Books.
TallBear, K. (2017). Beyond the life/not life binary: A feminist indigenous reading of cryopreservation, interspecies thinking and the new materialisms. In J. Radin, & E. Kowal (Eds.), *Cryopolitics: Frozen Life in a Melting World* (pp. 179–202). MIT Press.
Todd, Z. (2016). An Indigenous feminist's take on the ontological turn: "Ontology" is just another word for colonialism. *Journal of Historical Sociology, 29*(1), 4–22.
Truman, S. (2019). SF! Haraway's situated feminisms and speculative fabulations in English class. *Studies in Philosophy and Education, 38*(1), 31–42.
Wallin, J. J. (2015). Dark posthumanism, unthinking education, and ecology at the end of the Anthropocene. In N. Snaza, & J. Weaver (Eds.), *Posthumanism and educational research* (pp. 172–183). Routledge.
Weaver, J. (2010). *Educating the Posthuman*. Sense Publishing.
Whitehead, A. N. (1978). *Process and reality*. The Free Press.

Whyte, K. P. (2018). Indigenous science (fiction) for the Anthropocene: Ancestral dystopias and fantasies of climate change crises. *Environment and Planning E: Nature and Space*, *1*(1–2), 224–242.

Woodard, B. (2013). *On an ungrounded earth: Towards a new geophilosophy*. Punctum.

Young, E. B. (2013). *The Deleuze and Guattari dictionary*. Bloomsbury.

7

GENERATIVE OPENINGS

> There will always be some life if there is oxygen and water. Nature rises from the ashes. It will push on past the impossible.
>
> **Chiara, age 11, CC+Me Researcher**

Our conclusion consists of a series of generative openings that continue to carry the potentials of the Climate Change and Me (CC+Me) project into an uncertain and unknowable future. Rather than summarising the project as a complete and accountable body of work that lays claim to a particular academic territory, we linger in the thresholds held open by questions of "what next?" and "what else?" As Chiara elegantly summarises in the quotation above, this is also a question of how to "push on past the impossible" in cultivating and sustaining life ways that are adequate to the existential threat of climate change. As children and young people have continuously reminded us throughout our project and the writing of this book, the inhuman horror of climate change is something we need to get to know. It's a force we need to get close to. We need to get so close that we realise our inseparability from climate and the more-than-human forces of the Earth.

This closeness requires a certain ecological aesthetic. It requires a resistance to the nostalgic, sentimental, or reactionary view backwards, a view back to an earlier, easier, simpler aesthetic relationship with nature as a convenient background and resource for human progress. Instead of looking back on a failed attempt to mould nature to human ends, we need to collapse the perverse metaphysical distance between humanity and nature, even if that collapse necessitates an encounter with the possibility of human extinction. Many children and young people are already living the reality of this collapse, attempting to make sense of a world without distances, a world of vicarious dependencies and radically contingent mutualities of life.

DOI: 10.4324/9781003336006-7

Today's children and young people are now left with the task of learning how to live and die on a damaged and decaying planet. This requires a new role for education in the ashes of progressive modernity, including its ruined aspirations for the transcendence of "man" beyond the ecological dependencies that sustain life. What kind of education can bring us closer, rather than farther, to the joys and terrors of these dark ecological dependencies? Considered from this perspective, the imaginary is not a retreat or escape from the conditions of life, but is rather what presupposes the conditions of dependency which can be stretched, manipulated, mutated, curved, but never broken. To the extent that climate change alters the physical and metaphysical conditions which are the ground for life, it also alters the conditions of possibility for the creative expression of life's potentials. As Whitehead (1978) continuously reminds us, creativity is always *conditioned* by the stubborn fact of the world as an ever-changing matrix of experience. As much as humans are composed of "creatures of creativity" (p. 21), we are also creatures of the conditions through which we live and think. In this, we are akin to every other form of life, and indeed, every "puff of existence". Creatures are made of the world, and the world is made of creatures.

By focusing on generative openings rather than conclusions, we aim to sustain this book's commitment to the ongoingness of climate child imaginaries under the shifting conditions of creatures and worlds-in-the-making. Learning can never be separated from the act of *doing something*, of participating in the world's creative activity. This makes education indistinguishable from life-living, and especially, from art as a practice or "way" that embraces the fullness of life's creative potentials (Manning, 2016, 2020). In this spirit, our turn to generative openings is simply a gesture towards what's happening, where things are agitating, and how climate child imaginaries continue to resonate and evolve in the wake of the Climate Change and Me project.

Generative Openings into the Digital

Digital platforms were used creatively throughout Climate Change and Me, beginning with the design of the project website as a community hub for participating in the exchange of ideas, images, questions, imaginings, and conversations. Children and young people were encouraged to take on fictional avatars when participating in this digital space, opening their engagement with climate change beyond normative subject positions and identities. We were often surprised by the extent to which children and young people elected to use this digital space, which in some ways mirrored the functions of a social media platform, while also incorporating elements of multimodal blogging and sharing platforms for creative arts and writing.

The emergence of a collaborative experiment in building fictional characters and worlds, initiated by CC+Me researcher Jasmyne Foster, marked a turning point in the use of this digital platform as a kind of studio for speculative co-production. As discussed in Chapter 6, this led to the formation of fictional characters that took on their own digital life in the project, with the characters also moving from the digital realm into physical classrooms and other learning environments through the CC+Me Curriculum. These characters also came to feature in a creative online survey tool published on the project website, which assigns users a hand-drawn character depending on their responses to questions about climate change and its impacts.

In early 2019, we began a new project that involved the co-development of a climate education mobile application with young people, building on the digital innovations of Climate Change and Me. We engaged a group of 20 young people from across Australia as co-designers, building on our network of partnerships with youth-led environmental activism groups such as Australian Youth Climate Coalition and SEED Indigenous Youth Climate Network. In collaboration with Matt Cumming (digital developer), Maia Osborn (research fellow), and Thili Wijesinghe (research assistant), we worked with young people to co-design the Climate Action Adventure! application through a series of workshops at Southern Cross University, followed by regular online meetings via the Zoom platform. The app was launched in late 2021, and is available to download for free on both mobile devices and PCs at ccme.app.

The Climate Action Adventure! app offers an interactive platform to raise awareness and inspire meaningful climate action, combining elements of online gaming, science fiction, social media, and digital activism. The app is built around the science fiction story and characters designed by Jasmyne Foster, in which young people develop mutant abilities that help them adapt to the destructive impacts of climate change in the near future (Figure 7.1). In the Climate Action Adventure! app, you can choose to play as one of two fictional characters from this story:

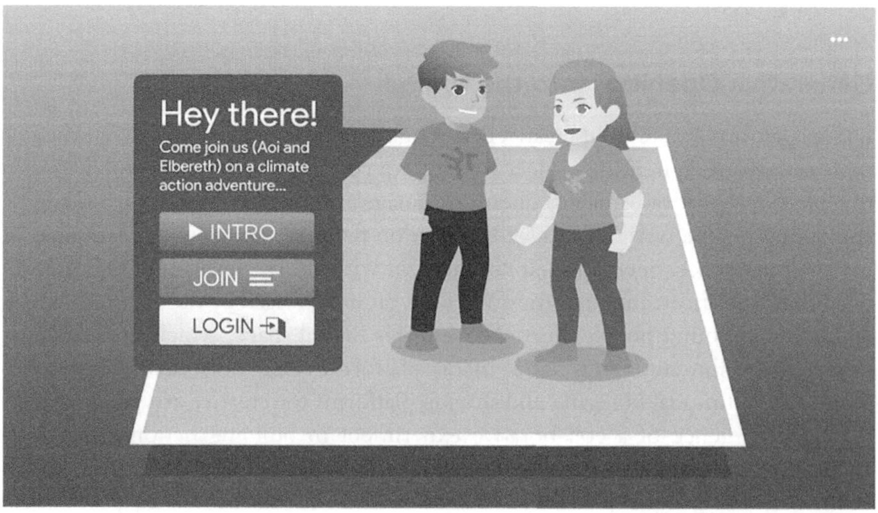

FIGURE 7.1 The design for the initial login screen of the Climate Action Adventure! App.

> **Aoi**, a young art student with a passion for climate science and environmental activism. After a near-death experience, Aoi realises that anything he draws in his notebook can come to life and help fight against climate injustice; or
>
> **Elbereth**, a young environmentalist with a passion for magic and the natural world. After helping bring Aoi back to life, Elbereth discovers that she can draw on elemental energies to heal people and other creatures affected by climate change.

Once you've selected a character, the app invites you into a virtual campus with its own cinema, library, art gallery, data dashboard, and live hashtag trackers for interacting with the latest climate science and activism campaigns (Figure 7.2). This dashboard connects data generated by user activity with climate science data and live social media updates sourced from all over the world. The "Top Tag Radar Report" continuously updates the most popular global hashtags connected to climate change activism, and allows users to link to an aggregated collection of top-rated posts across multiple social media platforms.

There is also an option to launch missions from the dashboard to explore how climate change is affecting the Earth's environments. Users can choose between four different "scapes" currently being impacted by climate change worldwide: Desertscapes, Forestscapes, Arcticscapes, and Cityscapes. As they explore each of these environments, users encounter "climate concerns" that have been crowd-sourced from children, young people, and adults all over the world (Figure 7.3). Each climate concern focuses on a particular species or issue within a larger ecosystem, how this is impacted by climate change, and what can be done to address this concern.

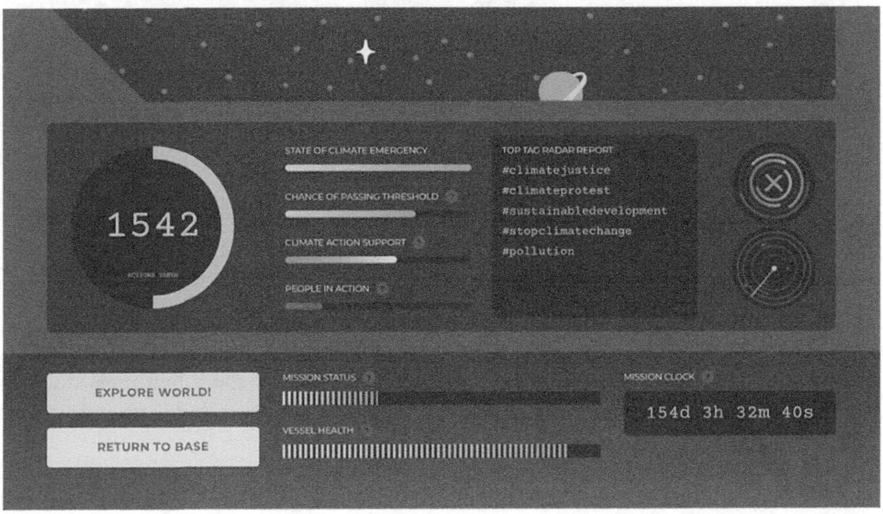

FIGURE 7.2 The Climate Action Adventure! dashboard, which provides a global overview of climate activism movements on social media, climate science, and user-generated activity data.

FIGURE 7.3 View from the Climate Explorer within the Climate Action Adventure! app, showing the option to explore the "Cityscape" which features urban climate concerns that have been crowd-sourced from children and young people all over the world.

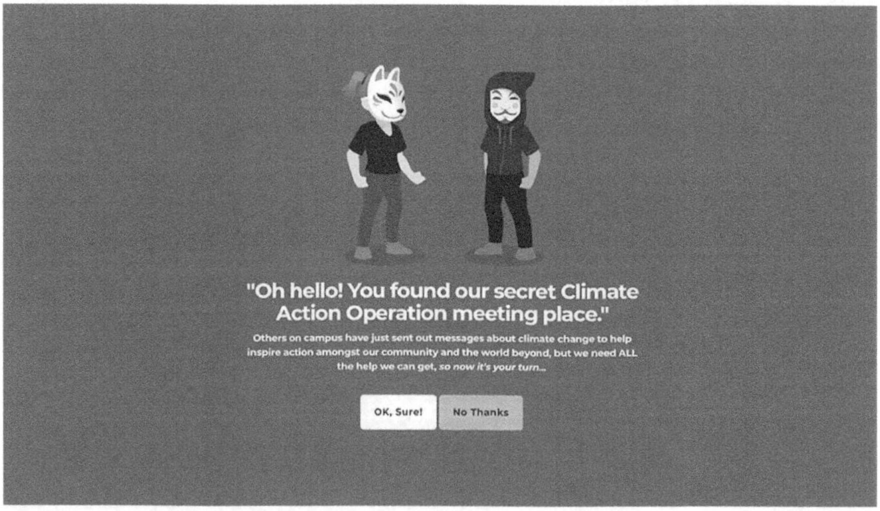

FIGURE 7.4 Screenshot from the Climate Action Adventure! app when users discover Aoi and Elbereth's secret activist hideout, which includes an anonymous direct feed into Twitter.

Arguably the most subversive element of our digital app design is Aoi and Elbereth's "secret hideout", which is hidden in the basement of an unmarked building within the app's virtual campus (Figure 7.4). This space enables users to generate their own hashtags and post anonymously into Twitter using masked versions of Aoi and Elbereth as their avatars. Inspired by the Atlantic Adaptation Theory platform designed by children in the original project, the app also links directly to a Discord server set up specifically for global discussions and engagement with climate change.

Combining elements of "hacktivism" with the character-building and world-building techniques of speculative fiction, these functions of the app connects climate child imaginaries with direct lines of climate activism in the sociopolitical domain. By masking the identities of young people and enabling them to post collectively in the guise of fictional personas, Aoi and Elbereth's hideout offers a generative opening for micropolitical activism through digital modalities of social re-imagining and intervention.

Generative Openings into the Environmental Arts

The environmental arts have also emerged as key areas of inquiry in the wake of Climate Change and Me, as crystallised through the international *Local Alternatives* initiative. *Local Alternatives* explores the transformative agency of environmental arts education in responding to the onset of climate change, biodiversity loss, and ubiquitous technology in urban spaces. Initiated during David Rousell's postdoctoral research fellowship in the Manifold Lab at Manchester Metropolitan University, *Local Alternatives* is conceived less as a project and more as a creative platform for intergenerational research collaborations committed to seeding diverse engagements with the environmental arts across international cities. As an agile hub for cross-generational learning and urban re-imagining, *Local Alternatives* is invested in the co-production of experimental projects through collaborations between children, young people, artists, designers, researchers, educators, curators, policymakers, and urban citizens from all walks of life. Examples of this work can be found on the *Local Alternatives* website: www.localalternatives.org.

These creative experiments are ongoing and open-ended, and often led by children and young people's sensory investigations, concerns, and imaginings of their local environments. Small projects are co-developed and co-produced from the ground up, with or without funding, with the aim of feeding into public exhibitions, social interventions, future projects, collaborations, and networks. By mapping and mobilising environmental arts initiatives across educational, cultural, and community sectors, *Local Alternatives* aims to shape urban policy and planning for more creative, inclusive, and sustainable cities, and provide an international hub for environmental arts research, practice, and knowledge exchange. As an ever-growing collective of collaborators and contributors, *Local Alternatives* has led to several significant exhibitions and events in galleries, museums, and public spaces across international cities, which to date have included events in Manchester, Berlin, Montreal, Edinburgh, Stockholm, and Melbourne.

A small project called Sensing Time involved a collaboration with the Young Contemporaries, a self-organising collective of young people who meet weekly at the Whitworth Art Gallery in Manchester, UK. Using an immersive exhibition by South African artist William Kentridge as a creative laboratory, the research team (David Rousell, Laura Trafi-Prats, Liz de Freitas, and Riikka Hohti) worked collaboratively with the Young Contemporaries to explore young people's shifting sense of time through the environmental arts. Responding to the onset of climate change and the urgency of decolonisation in conversation with Kentridge's work, the collective created an immersive exhibition called "Remixing Thick Time" in the grand hall of the Whitworth Art Gallery. The exhibition featured a series of interactive audio and visual installations, including a large-scale projection of collaboratively drawn animations; a pitch-black inflatable dome filled with the whispered voices of young people; and an interactive audio interface that invited audiences to "remix" sampled sounds and musical sequences collected by young people in the gallery.

Another small project evolved through a collaboration with Manchester Museum, a forward-thinking natural and cultural history museum with a 200-year history of colonial acquisition and scientific classification. Titled *Inheriting the Anthropocene*, this project sought to engage critically with the Anthropocene as a conceptual, material, and affective figuration of the current epoch, focusing on how museum spaces might be differently experienced and imagined by young people at the micropolitical level. Young people (ages 11–14) were invited to join the research team (David Rousell, Maggie MacLure, Riikka Hohti, and Hannah Chalk) through an open call for participation released through the Museum's network, social media, and flyers distributed across the city of Manchester. The call was phrased as an invitation for "movers, shakers, makers, hackers, activists, and dreamers" interested in disrupting and re-imagining the museum through environmental arts practices. This open-ended approach was distinctive in positioning the museum itself as a non-innocent space for young people to encounter, explore, cultivate, question, experiment, and express their shifting sense of inheritance in the Anthropocene epoch (see Rousell et al., 2021).

Over the course of this project, decolonial concerns came increasingly to the fore as the collective explored the complexity of young people's encounters with the museum's galleries, collections, and store rooms containing more than 4.5 million objects, the vast majority of which are never seen by the public. In this respect, Manchester Museum provided a site for grappling with the inherited effects of the epoch as both traumatic and enlivening for young people, while enabling our intergenerational collective to challenge a museological space that has historically conserved, or held in place, a particular (White, Western, male) image of the liberal, rational subject (Rousell et al., 2021). This work in the museum culminated in the co-creation of a public exhibit, featuring a Baroque "Cabinet of Curiosities" assembled from objects collected by young people from the museum's private storehouses; an interactive "Bureau of Unanswerable Questions", which contained hundreds of critical and incisive questions asked by young people as they explored the museum; and the "Blotwalk", an immersive re-mapping of the museum that marked colonial acquisitions, blind spots, and exclusions by installing abstract inkblots across the museum's galleries.

Generative Openings into Country

Our ongoing work in climate change education and research is also continuing to evolve in respectful conversation with Aboriginal and Torres Strait Islander cultures, philosophies, and communities. This constitutes a generative opening that foregrounds the voices and cultures of Indigenous young people in responding to the challenges of climate change, perspectives which have often been obscured or overlooked in current portrayals of climate education and activism. Culture is ineffably tied to environment and place, particularly with respect to Indigenous relationships with Country.

For many Aboriginal and Torres Strait Islander people, Country refers to much more than just the "land" or physical environment. This has significant implications for Indigenous research methodologies and for anyone interested in researching with Aboriginal and Torres Strait Islander people and communities (Somerville and Turner, 2020). Here we turn to the work of Bawaka Country collective (2013, 2018, 2020) in northeast Arnhem Land, an international collective of Indigenous (*Yolŋu*) and non-Indigenous (*ŋapaki*) researchers, artists, and community members.

> Country is an Australian Aboriginal English term used to refer to specific places or territories that are Aboriginal peoples' homelands. Australia is made up of many, many different Aboriginal nations, peoples and Countries. Country is rich and intensely meaningful. Country includes people, and it is also more-than-human as it includes winds, currents, animals, rocks, laws, custom, movement, song, knowledges, relationships, histories, presents, and futures. Country is alive, sentient and communicative, it encompasses tangible and intangible forces that connect, shape, create, mutually care and become together in, with and as place.
>
> *(Bawaka Country, 2018, p. 8)*

We are interested to learn from, with, and about the ways that climate change and Country are intricately entangled in the spiralling temporalities of Aboriginal relationships with place (Burarrwanga et al., 2019). The Bawaka Country collective (2020) share insights into these relationships through *Wukun*, a Yolŋu songspiral (often known as a "songline") from Bawaka Country that can be translated into English as the "Gathering of the Clouds" (p. 296).

> As the clouds gather, they are gathering people and people are gathering them. We are co-becoming. So that the co-emergence of people with each other and with weather and time speaks to a need to understand and attend to the weather and climate as us, to respond as weather rather than to it. Climate change is never over there or simply in the future. It is never in the past. It does not live in statistics on paper or models on a computer as if the paper/computer was an artefact that could have an existence separate from humans,

> from the trees and metals that grew it, from the humidity or the mites or from the ways these things all come into being together.
>
> *(Bawaka Country, 2020, p. 301)*

The Bawaka Country collective is exemplary in forging new practices and pathways for collaboration between Indigenous and non-Indigenous researchers through a relational ontology of co-becoming. Since the collective's formation in 2006, their work has been underpinned by co-developed protocols for intercultural collaborative research authored by, with, and through Country. These protocols are detailed in the *Intercultural Communication Handbook* co-authored by Bawaka Country collective (2018). This *Handbook* is described as "a co-production of knowledge in which Yolŋu authors, as well as Country itself, choose what to share and the terms under which this might be shared in order to advance their own work of intercultural teaching" (p. 8).

In our own work we are developing collaborations with Aboriginal Elders, mentors, and scholars to explore the intricate entanglements between climate change and Country (climate Country), and, specifically, the role of children, young people, and education within these entanglements. Through our work with Professor Tracey Bunda, a Ngugi Wakka Wakka Aboriginal woman and esteemed scholar of Indigenous studies, we are learning not only about Aboriginal protocols for intercultural engagements and knowledge practices with, as, and through Country, but also about our own settler genealogies and their entanglements with unceded First Nations lands. This involves learning to work as trusted allies in the development of transcultural and transgenerational lines of research, education, activism, and community work (Hernández et al., 2021; Williams et al., 2018).

Artful practices of listening, sensing, storying, and mapping are central to this work, as these offer creative potentials for developing a transcultural language and grammar of exchange that resists the colonising tendencies of the Western positivistic logic (Phillips and Bunda, 2018; Snepvangers, 2016). As we turn to the next series of openings in the unfolding of this collaborative work, we hope to better understand the intricate relations of climate and Country through and with Indigenous philosophy, including through relational notions of research as *co-becoming* (Bawaka Collective, 2018), as *ceremony* (Wilson, 2008), and as *storying* (Phillips and Bunda, 2018). These Indigenous concepts of relationality are potentially transformative, with the understanding that colonising terms of "connection", "networking", "community-building", and "change" can be significantly retheorised through Indigenous perspectives, particularly when the entangled natures of Indigenous relationships with Country are foregrounded (Hernández et al., 2021). This also opens alternative possibilities for art and aesthetics as critical platforms for the community mobilisation of climate adaptation and mitigation strategies with Indigenous and non-Indigenous young people.

As we continue to face the challenges of a precarious and contested past, present, and future, we hope to bring Indigenous and non-Indigenous young people's relations and knowledges of Country together in transformative ways. We see the potential for children and young people to lead transcultural responses to climate change across the diversity of Australia's wild, rural, regional, and urban cultures and environments. Rather than concluding with a closing, we linger with an opening.

References

Bawaka Country including Burarrwanga L., Ganambarr R., Ganambarr-Stubbs M., Ganambarr B., Maymuru D., Suchet-Pearson S., Wright S.L., Lloyd K., Tofa M., Daley L. (2018). *Intercultural communication handbook*. http://bawakacollective.com/handbook/.

Bawaka Country including Suchet-Pearson, S., Wright, S., Lloyd, K., Burarrwanga, L., & Bawaka Country. (2013). Caring as Country: Towards an ontology of co-becoming in natural resource management. *Asia Pacific Viewpoint, 54*(2), 185–197.

Bawaka Country including Wright, S., Suchet-Pearson, S., Lloyd, K., Burarrwanga, L., Ganambarr, R., Ganambarr-Stubbs, M. … & Maymuru, D. (2020). Gathering of the clouds: Attending to indigenous understandings of time and climate through songspirals. *Geoforum, 108*, 295–304.

Burarrwanga, L., Ganambarr, R., Ganambarr-Stubbs, M., et al. (2019). *Songspirals: Sharing women's wisdom of country through songlines*. Allen & Unwin.

Hernández, K. J., Rubis, J. M., Theriault, N., Todd, Z., Mitchell, A., Country, B., … & Wright, S. (2021). The creatures collective: Manifestings. *Environment and Planning E: Nature and Space, 4*(3), 838–863.

Manning, E. (2016). *The minor gesture*. Duke University Press.

Manning, E. (2020). *For a pragmatics of the useless*. Duke University Press.

Phillips, L., & Bunda, T. (2018). *Research through, with and as storying*. Routledge.

Rousell, D., Hohti, R., MacLure, M., & Chalk, H. L. (2021). Blots on the Anthropocene: Micropolitical interventions with young people in a university museum. *Cultural Studies ↔ Critical Methodologies, 21*(1), 27–40.

Snepvangers, K. (2016). Bending the twig: Indigenous perspectives in tertiary art and design. *Australian Art Education, 37*(2), 165–183.

Somerville, W., & Turner, B. (2020). Engaging with indigenous research methodologies: The centrality of country, positionality and community need. *Journal of Australian Studies, 44*(2), 182–184.

Whitehead, A.N. (1978). *Process and reality*. The Free Press.

Williams, L., Bunda, T., Claxton, N., & MacKinnon, I. (2018). A global de-colonial praxis of sustainability – Undoing epistemic violences between indigenous peoples and those no longer indigenous to place. *The Australian Journal of Indigenous Education, 47*(1), 41–53.

Wilson, S. (2008). *Research as ceremony: Indigenous research methods*. Fernwood Publishing.

INDEX

Pages in *italics* refer figures and pages followed by n refer notes.

For Product Safety Concerns and Information please contact our EU representative GPSR@taylorandfrancis.com Taylor & Francis Verlag GmbH, Kaufingerstraße 24, 80331 München, Germany